D0228285

Practical Construction Management

Also available from Taylor & Francis

Construction Project Management
Peter Fewings

Taylor & Francis   HB: 0-415-35905-8
PB: 0-415-35906-6

Building Regulations Explained
John Stephenson, London District Surveyors Association

Taylor & Francis   HB: 0-415-30862-3

Understanding Building Regulations, 3rd edition
Simon Polley

Taylor & Francis   HB: 0-415-34917-6

Human Resource Management in Construction Projects
Martin Loosemore, Andrew Dainty, Helen Lingard

Spon Press   HB: 0-415-26163-5
PB: 0-415-26164-3

Writing Engineering Specifications, 2nd edition
Paul Fitchett, Jeremy Haslam

Spon Press   HB: 0-415-26302-6
PB: 0-415-26303-4

Information and ordering details

For price availability and ordering visitour website www.sponpress.com
Alternatively, our books are available from all good bookshops.

Practical Construction Management

R. H. B. Ranns with E. J. M. Ranns

Taylor & Francis
Taylor & Francis Group
LONDON AND NEW YORK

First published 2005
by Taylor & Francis
2 Park Square, Milton Park, Abingdon, Oxon OX14 4RN

Simultaneously published in the USA and Canada
by Taylor & Francis
270 Madison Avenue, New York, NY 10016

Taylor & Francis is an imprint of the Taylor & Francis Group

© 2005 R. H. B. Ranns and E. J. M. Ranns

Printed and bound in Great Britain by
TJ International Ltd, Padstow, Cornwall

Publisher's Note
This book has been produced from camera-ready copy supplied by
the authors.

British Library Cataloguing in Publication Data
A catalogue record for this book is available from the British Library

Library of Congress Cataloging-in-Publication Data
Ranns. R. H. B.
 Practical construction management / Ray H.B. Ranns & Edward J.M.
Ranns.
  p. cm.
 ISBN 0-415-36257-1 (pbk.)
 1. Building--Superintendence. 2. Contractors. 3. Subcontracting. I. Title.
TH438.R3356 1990
690'.068--dc22

2005013680

ISBN 0-415-36257-1

Contents

Introduction

This book builds on "The Site Agent's Handbook", to both update and widen its application to Main and Subcontractors operating in Civil Engineering, Construction and Process as well as to make it relevant to an international market. Whereas the legal and business framework is different around the world, the management disciplines are identical. The major changes in the Construction Industry in the UK since the publication of The Site Agent's Handbook in 1990 have been:

- The Latham and Egan Reviews leading to a more collaborative approach;
- the inception and rise in the use of the New Engineering Contract which embraced these principles and includes Adjudication;
- International Management Standards for Quality Procedures and Environmental Matters;
- statutory right to Adjudication with an enforceable decision in 28 days from referral within the UK
- the Construction Design and Management (CDM) regulations prescribing and codifying the system to design and manage construction.

The latter item is not applicable outside the UK but the procedures within the CDM Regulations are a useful standard and would no doubt serve as protection against litigation if they had been followed.

The central theme of this book is to provide practical advice on the proper control of resources, financial control and contractual dealings with other organisations. The book has been designed as a series of standalone sections that can be consulted separately to provide advice on any particular problem.

Universities and Colleges are now including more management and contractual topics in Civil Engineering and Building degrees, but of necessity, these tend to be more headline principles rather than the basics of daily construction disciplines. Daily Construction Management is not a jungle and, if the disciplines and contractual rules are absorbed and applied, it is no different than solving mathematical problems.

There is a growing tendency in contracting companies to split the operational and financial management at site level. A central theme in this book is the responsibility of all of the Contractor's team in the financial success of each project and the need to mesh the production and financial disciplines on site into an effective team. Where operational management ignores the financial and contractual aspects and a true team is not created the end results can be very adversely affected.

It is important to place the principles and guidelines within this book in context of the Industry as a whole and Chapter 1 on the Background should not be missed.

The order of the chapters has been deliberately chosen thereafter starting with Chapter 2 'The Tender' because the tender contains all the assumptions and base productivity data that a Site Manager needs to manage any project. The natural successor to this is to consider the disciplines and actions required when starting up a project and these are taken in Chapter 3.

The purpose of any Contractor is to generate profit and this is achieved by rigorous cost control which is particularised in Chapter 4. Contributing to this aim is the successful management of Subcontractors and labour which are taken in Chapters 5 and 6 and the backstop provisions of insurance are taken as well. After the Contractor or Subcontractor's management of their resources, the biggest variable is his contractual liabilities to his Client. The basis of a contract and the key contractual actions for Site Management are taken in Chapter 8.

A centre piece of the book is the Contract Appendix, the main purpose of which is for a user of one form of contract who finds himself with a different form to be able to check the differences. Thus practical contractual guidance is given across a range of contracts, the Contract Appendix now compares and contrasts.

- The ICE Conditions 5th, 6th and 7th Editions and Design and Construct
- The Highways Conditions
- FIDIC
- The NEC 2nd Edition (not expected to significantly affected by NEC 3rd)

The Irish 3rd Edition follows the ICE 5th Edition and the 4th follows the 6th. Users of the ICE Term Contract are also catered for as this follows ICE 7th Edition. The analysis is also applicable in Ground Investigation: the ICE Ground Investigation Conditions, 1st Edition follows the ICE 5th and the 2nd Edition the ICE 7th.

The payment of extras and their effects on a project are not always readily agreed, the disciplines necessary when making a claim are contained in Chapter 9. Statutory Adjudication in the UK has meant that more disputes are being taken to Adjudication involving site based personnel during the course of the project and the procedures and requirements have been considered in Chapter 10.

The CDM Regulations affect the duties of all organisations involved in construction within the UK and provide a template for good practice internationally. This subject has been allied with Design Management which is increasingly within the Contractor's domain with the rise of Design and Construct contracts and is taken in Chapter 11.

Quality, environmental, and health and safety systems have been taken last, in Chapter 12, not because of the importance assigned to them, but because systems that comply with international standards are not universally employed and vary considerably in detail between companies and are not in the powers of Site Management to install.

CHAPTER 1

Background

1.1 THE PURPOSE OF THE BOOK

This handbook is aimed at Construction Managers and therefore aims to provide the reader with a guide to the actions necessary for the Site Team once a contract has been entered into.

It does not dwell on whether a particular form of contract or method of construction is preferable because this has generally been decided by the Employer before the contract was entered into. The key objectives are therefore:

- how to control costs both in terms of utilisation of resource and continuous monitoring of financial outcome;
- how to understand and enforce the bargain that has been made with the Employer or Client;
- how to make the bargain with a Subcontractor and then hold them to it:
- how to plan the crucial start up of a project;
- how to control quality;
- how to comply with regulations.

The emphasis on enforcement of the bargains made, otherwise known as Contracts, at first sight seems confrontational in the current climate for improved relations and the buzz words of teambuilding and partnering and hence it is as well to review current developments and set the detail in context.

1.2 DEVELOPMENTS AND THE DIRECTION OF CHANGE IN THE INDUSTRY

There have been many initiatives worldwide to improve productivity and reduce construction times. In the UK there have been two major reviews which can be summarised and set in context as follows.

The first was "Constructing the Team" in 1994 lead by Sir Michael Latham which still holds important messages for the industry in terms of developing Best Practice through promoting the search for:

- value for money;
- better productivity;
- better competitiveness of the industry.

Latham recognised that contractual relationships were impeding progress towards these aims and laid down a series of 13 principles that he maintained should be present in all construction contracts.

These were as follows:

1. A specific duty for all parties to deal fairly with each other, and with their subcontractors, specialists and suppliers, in an atmosphere of mutual co-operation.
2. Firm duties of teamwork, with shared financial motivation to pursue those objectives. These should involve a general presumption to achieve "win-win" solutions to problems which may arise during the course of the project.
3. A wholly interrelated package of documents which clearly defines the roles and duties of all involved, and which is suitable for all types of project and for any procurement route.
4. Easily comprehensible language and with guidance notes attached.
5. Separation of the roles of contract administrator, project or lead manager and adjudicator. The project or lead manager should be clearly defined as client's representative.
6. A choice of allocation of risks, to be decided as appropriate to each project but then allocated to the party best able to manage, estimate and carry the risk.
7. Taking all reasonable steps to avoid changes to pre-planned works information. But, where variations do occur, they should be priced in advance, with provision for independent adjudication if agreement cannot be reached.
8. Express provision for assessing interim payments by methods other than monthly valuation, i.e. milestones, activity schedules or payment schedules. Such arrangements must also be reflected in the related subcontract documentation. The eventual aim should be to phase out the traditional system of monthly measurement or remeasurement but meanwhile provision should still be made for it.
9. Clearly setting out the period within which interim payments must be made to all participants in the process, failing which they will have an automatic right to compensation, involving payment of interest at a sufficiently heavy rate to deter slow payment.
10. Providing for secure trust fund routes of payment.
11. While taking all possible steps to avoid conflict on site, providing for speedy dispute resolution if any conflict arises, by a predetermined impartial adjudicator /referee/ expert.

12. Providing for incentives for exceptional performance.
13. Making provision where appropriate for advance mobilisation payments (if necessary, bonded) to contractors and subcontractors, including payments in respect of off-site prefabricated materials provided by part of the construction team.

Only the NEC contract came close to these aims and the second edition 1995 (also called the ECC) was brought out to achieve all these objectives. Needless to say, having commissioned the report the NEC contract is still not the Government's usual contract when it undertakes major projects.

The next significant event was on the international stage, being the introduction of ISO 9000 in 1996 relating to Quality Management (see Chapter 12) which codified methods for providing continuous improvement in Quality Management.

The next UK Government-sponsored initiative was "Rethinking Construction" in 1998 lead by Sir John Egan. This built on the 1994 Latham Report and the ideas within ISO 9000 and took them further, clearly defining areas for improvements and making specific recommendations.

The practical goal set by Egan was:

"Our targets are based on our own experience and evidence that we have obtained from projects in the UK and overseas. Our targets include annual reductions of 10% in construction cost and construction time. We also propose that defects in projects should be reduced by 20% per year"

The five key "Drivers of change" identified as essential elements in achieving these goals were:

1. Committed leadership.
2. A focus on the customer.
3. Integrated processes and teams.
4. A quality driver agenda.
5. A commitment to people.

These objectives are settled at Board level and govern the way in which Clients seek to have facilities constructed and a contracting company seeks to do business and enter into contracts. The proposed drivers for change are not something which can be expected to alter contracts once they are let and are not in the province of everyday construction managers.

This is not the case with the series of techniques which were proposed to facilitate the recommendations that are appearing in one form or other in many projects.

1. **Value Management**: eliminating waste from the brief and ultimately the design.

2. **Benchmarking**: understanding and measuring performance. Setting improvement targets.
3. **Cultural changes**: principles of "zero defects".
4. **Integrated processes**: Utilising the full skills of the construction team to deliver value to the client.
5. **Product development:** a commitment to develop a generic product by innovation to meet and exceed the needs of the client.
6. **Project implementation**: develop the generic product into a specific project or site.
7. **Partnering the supply**: driving innovation to establish sustained chain improvement up and down the supply chain. To share in the rewards of improved performance.
8. **Production components:** design and development of a range of standard components.
9. **Lean thinking:** eliminate waste in the production process, to increase the value.
10. **Sustained performance**: continuously adding value and maintaining improved efficiencies.

To achieve these goals, relationships have to last for longer than one project between the Client and his Design Team (if not lead by the Contractor) and then between the Contractor, his designers, subcontractor and suppliers. These again are not part of upholding the bargains made on a particular project.

There is a cultural change necessary for clients who have to be weaned off the cheapest initial price and opt for more collaborative arrangements. Success is being achieved where clients are constructing similar projects over many sites, for instance:

1. Tesco supermarkets have excellent arrangements and continually improving results from using a few selected contractors on a continuous programme.
2. The Highways Agency has developed an "Early Contractor Involvement" scheme and now uses a form of Design and Construct contract on award.
3. Road maintenance contracts are being let on an area basis for several years.
4. McDonald's "Drive Through Restaurants" are almost entirely prefabricated, reducing site works to a minimum.

There are fewer improvements in the housing sector where prefabrication meets with customer resistance.

The second cultural change is an end to the separation of design and construction. This may sound easy but design professionals are normally contacted first when a potential first time construction employer is seeking advice. The natural instinct for the professional is to want to keep all the design and immediately creates the traditional separation. A feasibility study with fees for a Design and Build Contract is far less lucrative to the professional.

Contractual arrangements which novate the Designer's contract to the Contractor do not achieve the results sought, although they do reduce the Employer's risk.

However, there is a reverse flow in some sectors, including public bodies, where there is a growing use of reverse internet auctions, in some cases open to any contractor, subcontractor or supplier that goes against these principles.

Taking the matters that can be applied to single projects there are some initiatives that can apply. Even on single projects there is a growing tendency to have a separate partnering agreement. These are even added to contracts which are fixed price where there is no share in the rewards.

Partnering agreements are of two types which split between parallel "charter" expressing an agreement to co-operate which states that it has no effect on the contractual arrangements. They provide set forums for discussion of contentious issues and do encourage co-operation, which is an obvious benefit and should always have been a priority.

One of the few contracts with a specific contractual partnering option is the NEC Option X12, but even here it states in the introduction:

> "The Partnering Option does not include direct remedies between non-contracting Partners to recover losses suffered by one of them caused by a failure of the other. These remedies remain available in each Partner's own Contract, but their existence will encourage the parties to compromise any differences that arise."

It remains vital, even when in partnering relationships or any other form of contract, to adhere to the required procedures under the formal contract.

Value Engineering Workshops are becoming more common even on traditional client-designed contracts where the Contractor, his Subcontractors and Suppliers make suggestions that will improve buildability and reduce programme and substitute specifications that can reduce cost.

Lean Construction has had considerable exposure in the technical press, but is not a panacea. The principles were established in manufacturing; construction projects as a whole, despite areas of improvement, are still one of a kind, executed on site (rather than prefabricated in a series of factories) by a temporary organisation (in that the team is assembled for the project).

Much of what proponents claim for lean construction is a restatement of the Egan Principles. The Chartered Institute of Building claims on its website:

> "Applied to construction, Lean techniques change the way work is done throughout the different stages of a project. Lean Construction following the objectives of a lean production system (to maximise value and minimise waste) to specific techniques and applies them in a new project delivery process.

The results are:

- better integration between design and construction processes with active role of customers;
- a phase-by-phase structured work to maximise value and to reduce waste at the project delivery interfaces
- performance Management aims at improving total project performance because it is more important than reducing the cost or increasing the speed of any activity.
- "Control" is redefined from "monitoring results" to "making things happen". The performance of the planning and control systems is measured and improved.

The reliable release of work between specialists in design, supply and assembly assures value is delivered to the customer and waste is reduced. Lean Construction is particularly useful on complex, uncertain and quick projects. It challenges the belief that there must always be a trade between time, cost, and quality. Some key terms in Lean Thinking are:

- dependence (influence of preceding tasks);
- variation (due to presence of uncertainties);
- buffer (mechanism to deaden the impact of a variation);
- throughput (the output rate of a production process);
- point speed (the rate of work in a sequence at a particular point);
- capacity (capability)".

There is nothing in this explanation of Lean Construction that is not already covered by the generality of the Egan proposals or is other than the timely control and forward planning using computer networks described in Chapter 4: "Cost Control" and IS0 9000 Quality Systems described in Chapter 12.

Major construction projects always had to employ sophisticated planning techniques combined with knowledge of the project. It may be that manufacturing came late to this understanding and has recycled the principles in new jargon.

1.3   THE THREE INTERLOCKING ELEMENTS OF CONTRACTING

Before considering the rigorous discipline to contain and control cost and to enforce bargains that have been made it is as well to look at the totality of construction.

There are three interlocking aspects of any construction company's operations. These are often expressed diagrammatically as three circles which interlock, labelled "Getting Work", "Doing Work" and "Getting Paid". Each has a profound effect on the other and is essential to the survival of the organisation.

If work is not obtained it is impossible to continue. However, if work has not been done properly in the past, i.e. to time, budget and quality, then that reputation will make the getting of work more difficult.

Similarly, a reputation for pursuing every penny and a marked tendency for adjudication, litigation and/or arbitration will reduce the opportunities for tendering. If work has been done to time and the required quality then the process of getting paid is easier.

1.3.1   Marketing or Obtaining Work

Public Sector for Local Authorities, Government departments and other public bodies maintain standing lists of suitable contractors for smaller contracts. On any scheme over £600,000 they are currently required to advertise in the European Journal for interested contractors. The Employer on smaller projects will normally use all or some of the following broad criteria to establish his standing lists and any particular tender list:

1. Financial standing of Contractor
2. Plant at his disposal
3. Number of suitably qualified staff and labour
4. Safety record
5. Number of recent similar projects successfully completed
6. Size of regional or local input
7. Commitment to quality initiatives
8. Number of arbitrations or litigations in progress.

Contractors endeavour to persuade, by detailed submission and meetings with prospective employers, their particular expertise and suitability for inclusion on set tender lists. Prequalification submissions for the larger public sector and utilities projects in some areas are now costing as much as tendering for the resultant work.

Public sector partnerships are starting to appear but the criteria for inclusion are basically the same. Private sector work does not always have to be advertised. It is in this field that local knowledge makes it possible for a short tender list to be settled before it is generally known that a project is to be constructed.

Subcontractors for significant or key elements of the larger schemes are also being involved in formal prequalification documents to insert in the Contractor's submissions.

The bulk of a subcontractor's marketing effort is directed to Main Contractors to develop long term relationships over several projects. Where this is not achievable the secondary objective is to ensure that they are on the lists of subcontractors held for each category of work in both the head and area offices of all contractors so that they have an opportunity to price all relevant schemes.

1.3.2   Contract Execution

This Handbook excludes detailed consideration of construction methods, specification compliance, and production data for plant and equipment, staffing and overhead levels.

As a new reader studies the chapters on contractual matters and payment the essential point to remember is that the Contract is the means by which the risks and rewards of construction are divided between the Employer and the Contractor. On most forms of contract the Engineer/Architect/Supervising Officer stands between them as a quasi-arbitrator.

The Employer can reduce the risks in size and number by the amount of pre-tender work done in design, drawings, comprehensive bills of quantities, even early contractor involvement and, probably most importantly, the extent of site investigation. All of these items, particularly the last, are expensive and there is obviously a point at which further work brings no comparable savings.

Employers still tend to look at the Contract Sum or total of the Bills of Quantities as the total they are going to pay for the project. This attitude puts pressure on the Engineer/Architect/Supervising Officer under the traditional form of contract to resist all additional payments particularly when combined with Employers trying to seek recovery of extras due to faults in design or information supply from their professional indemnity insurance or fee payments.

Many larger public contracts in the UK are moving away from this scenario and using Design and Construct-based contracts which are a development of an Indicative Design which has passed through Planning Control. The sum is fixed or is a series of fixed sums and with detailed design in the hands of the Contractor and his Designer the final outturn for the Employer is more certain. There has also been an increase in Target Cost-based variants where the share mechanism provides an incentive to the Contractor to reduce the end costs through value engineering.

Although alternative forms of contract may be relevant for major projects, the traditional methods remain the most common for the majority of projects.

1.3.3   Getting Paid

The principles are the same whether a contract is one of a series under a long-term partnership arrangement or a one-off traditional contract. Knowledge of Tender Assumptions and Methods is central to getting paid the correct contractual amounts and making recoveries from insurers or subcontractors.

The chart gives the intrinsic thought process that a site team should go through as soon as it thinks expenditure is exceeding income.

It can either try to recover more money from the Client (Area A), reorganise his site (Areas B and C), or recover more from subcontractors (Area D).

When allocating anything into Area C the Site Team must take the necessary action to remedy site based faults.

The following chapters cover the need to protect the assumptions made at tender. The Directors implicitly approve those assumptions and it will be foolhardy for the Site Team to take what might at first sight seem to be the easy option of assuming an error or inadequacy in the tender. Chapters 7, 8 and 9 give the means by which recovery is made.

CHART DEMONSTRATING THE THOUGHT PROCESS TO BE FOLLOWED BY THE SITE TEAM

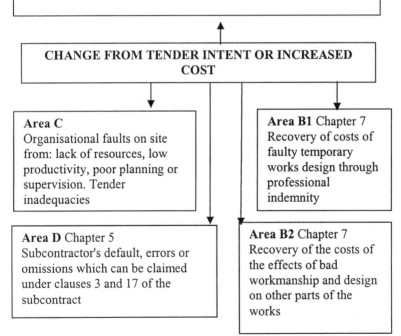

Area A
Recovery from the Client as a result of circumstances which match the criteria described in Chapter 8 and by methods described in Chapter 9. Namely, late or inaccurate drawings, late approval of programmes, varied methods, varied sequences, instructions, variations, unforeseen circumstances inter alia. (Detailed in Chapters 8 and 9). Also recovery for damage covered by excepted risks.

CHANGE FROM TENDER INTENT OR INCREASED COST

Area C
Organisational faults on site from: lack of resources, low productivity, poor planning or supervision. Tender inadequacies

Area B1 Chapter 7
Recovery of costs of faulty temporary works design through professional indemnity

Area D Chapter 5
Subcontractor's default, errors or omissions which can be claimed under clauses 3 and 17 of the subcontract

Area B2 Chapter 7
Recovery of the costs of the effects of bad workmanship and design on other parts of the works

CHAPTER 2

The Tender

2.1 INTRODUCTION

Throughout the execution of the project the assumptions in the tender must always be the guiding light to the Site Team, as detailed in the diagram on page 11. The price for any item or operation is built up from a series of assumptions of conditions, resources, productivity and methods and each links to the other operations. Proposed changes to tender methods should be carefully considered as to whether they are more cost effective and do not affect the overall planned resource balance. If a change is necessary which comes about because the tender assumptions are no longer valid the answers to the following questions must be sought:

1. were the assumptions reasonable at the time of tender?
2. if the answer to 1. is yes, then any additional costs involved must be recoverable from some source (see Chapter 9 "Claims").
3. if the answer to 1. is no, there is an error, that error might be included under design of temporary works which may be recoverable under Professional Indemnity Insurance which is discussed later.

Throughout the Site Team's dealings with Subcontractors, the Client and Engineer, they must at all times, for maximum efficiency and to give backing to their actions, be aware of the construction of the tender, the assumptions made and the spread of money.

To understand the tender the Site Team must appreciate the way in which the majority of Main Contractors arrive at their tender and the pressures which cause it to become an imperfect document so that the Contractor's costs are not reflected in the Bill of Quantities descriptions, individual Activity Schedules or Milestones.

The majority of Main Contractors use analytical five column estimating based on computer spreadsheets. There are still efficiency gains if employers can be persuaded to issue and accept electronic bills of quantities, where these are used, and receive an electronic tender. This appears to be some way off yet for security rather than technical reasons.

Although full analytical estimating is not necessary for a small drainage subcontractor restricting himself to standard estate drainage, for bigger organisations it is essential. For example, the Senior Estimator of a £15 million turnover company, which was in receivership, was once asked about the build up of site overheads on a particular contract and the profit margin expected. It transpired that their method of tendering was to write rates against the quantities and to rely on preliminaries and the balancing item to provide finance at the start of the contract. The company had no idea what profit margin was available within its tender. They took several jobs too cheaply with the result that they went out of business.

2.2   FIVE COLUMN ESTIMATING

When the contract does not include Bills of Quantities the Contractor has to execute a take off to allow him to accurately price the work required. The Contractor will probably take off the major items only and allow for the minor items in a general uplift in the allowances rather than incur the cost of a detailed take off in accordance with a standard Method of Measurement.

The five columns used are labour, plant, materials, subcontractors and Provisional and PC sums. These are either ruled on every page of one copy of the bill or automatically available on a computer estimating programme or standard spread sheet. The estimator then calculates the labour, plant and material elements as described in the pricing operation and inserts the subcontract quotes. The PC and Provisional sums are entered into the fifth column. The rate for each bill item is obtained by adding laterally, except when it is decided to spread monies through the bill as is described later under the heading "Completion of the Form of Tender and Bill of Quantities".

The individual columns of labour, plant, materials and subcontractors are added up page by page, providing a split for each section and eventually a split for the entire measured work section. All companies have a series of standard preliminaries sheets which form an aide memoir to the Estimator so that he prices every item. They would cover the following areas:

1. A Summary Sheet on which the five section totals appear together with the total of the preliminaries to which the overhead and profit additions are made
2. Site staff
3. Offices and sheds
4. Temporary services
5. The Client side accommodation requirements
6. Freight and site transport
7. Equipment
8. Clearance and maintenance
9. Sundries

10. Scaffolding
11. Plant not in rates
12. Temporary works
13. Traffic management
14 Service gangs
15. Insurance
16. Bonds
17. Design fees
18. Head or Group Office overheads
19. Subcontractor's summary
    (This lists all subcontractors approached and their prices, showing the spread of prices and giving an idea of the risk.)

2.3   THE OPERATIONS PERFORMED BY THE ESTIMATING TEAM PRIOR TO PRICING

It is a feature of modern times, with all public expenditure geared to yearly financial targets, that project construction times are getting shorter and the time allowed for preparing the tender is also becoming shorter. Typically, for all but the largest projects, only four weeks will be allowed. In this analysis it is assumed that the Estimator is going to do, or direct, all the tasks with clerical assistance. However, in many larger firms the task is divided between buying, design, planning and quantity surveying departments.

To make best use of the time available the first operation is to "strip out" the bill. This involves producing enquiries for major material items and all intended subcontractors. It has the distinct disadvantage that the programme and method statement have not yet been considered and will not have been included in the subcontract enquiries. As emphasised in Chapter 5, the Site Team must watch for compatibility of quotation conditions with the contract programme and method statement.

The Estimator next reads in detail all the tender documents and in particular watches for the following:

1. The allocation of risk in the Main Contract Form (see Contract Appendix)
2. Additional and non-standard Contract and Specification clauses
3. The preamble to the Bill of Quantities where any non-standard measurement clauses will be included
4. Any additional or varied clauses to the standard form of contract. He then makes full check of the main significant quantities against the drawings to establish the main work areas.

This is now the best time for the all important site visit when the salient points are in the Estimator's mind and he can visually check the awkward corners and access problems.

Simultaneous considerations will now be given to the tender programme and the temporary works. These tasks can be carried out by specialists or the Estimator himself, depending on the company and the project. Unless it is a specific requirement of the contract, the tender programme is not submitted with the bid. If the particular documents call for its inclusion then the comments on contract programmes under Chapters 3 (Starting up a Project) and 8 (Contract) should be borne in mind and added care given to the preparation.

The object of tendering for work is to win it. Therefore successful estimators have to be logically optimistic. The Estimator bases his assumptions on the logical interpretation of the documents and information readily available from the site visit. The contract divides the risk and rewards between the Employer and Contractor (see Chapter 8) and the Estimator will be wasting his time if he introduces any additional potentially adverse assumptions.

One very experienced Resident Engineer once said to me that, "You Contractors must foresee that there is going to be some unforeseen circumstances." He was right in that they nearly always occur. This is a contentious issue but every contract form identifies, classifies and divides the risks, the only caveat being the assumption that the Contractor is experienced. It is the Employer who should allow for them in his budget.

The Estimator, in preparing the programme, will be trying to plan logically to complete the contract in the shortest time in order to save on the preliminary costs. The sequences and method used must be logical and avoid excessive resource peaks either for the Main Contractor's own work or the subcontract works. Design of temporary works must also follow the principles in the preceding paragraph. They must be complete in time for the Estimator to price realistically and split their cost between fixed cost and time-related costs.

The resulting programme for major contracts will be fully resourced with all operations shown, so that a summary at the bottom will give a histogram of labour, plant and staff resources.

2.4   THE PRICING OPERATION

The Estimator will either build up a labour rate for the Contractor's own men, recognising anticipated travelling time, bonus etc., or he will assume a labour only rate for the particular location from a Labour Only Subcontractor. Then using a company plant schedule or hire list he will build up gang costs.

Using drainage for an example the following might be derived, assuming that trench sheets and the like are taken in the temporary works' preliminaries.

	Labour	Plant
1 No. Ganger £10.00 hr	£10.00	
2 No. Labourers £9.00 hr	£18.00	
360° Digger £15.00 hr		£15.00
Dumper £3.00 hr		£3
2 No. Drivers £10.00 hr	£20.00	
	£48.00/hr	£18/hr

Having derived these gang costs the Estimator will decide on the outputs. Let us assume that for a 225 diameter drainage at 5 metres depth he takes output at 1.5 LM per day and at two metre depth 5 LM per 8-hr day in the expected ground conditions. Then the labour and plant columns would read as follows:-

	Labour	Plant
225 diameter drains average depth 5 metres	£256	£96
225 diameter drains average depth 2 metres	£76.8	£28.8

He will then have to calculate the cost of the material surrounding the pipe, including a factor for wastage, and add to it the linear cost of the pipe. He might then include under the subcontractor column an amount for disposing to tip of surplus arisings, if the earthworks were being subcontracted and the Estimator was assuming the earthworks subcontractor would do this work. The labour and plant build ups will give the Site Team the benchmark against which to judge their actions on the subsequent contract.

As his last major operation the Estimator puts in the best subcontract prices received to that time. The Estimator completes the task by taking from his resourced programme all the preliminary items and costing them. All the information now exists to derive an initial tender figure. The next exercise is to provide, possibly against the elemental totals, the fixed price adjustment based on expected staff rises, labour annual awards and material price rises. These last items are normally based on the published projected rate of inflation. Individual consideration must be given to specialist items not included in published projected rates.

The remaining time before the tender review is spent by the Estimator trying to improve upon his external quotations and doing a cross check to see that the financial summation of the resourced programme and the estimate total correlate.

2.5   THE TENDER REVIEW

Up to tender review the objective has been to produce an accurate estimate of the cost to the company for executing the works. What now follows is a combination of checking and commercial decisions which turn an estimate of cost into a commercial bid or tender. Whether the review is carried out at a single meeting or in stages will depend on the company and the value of the tender. It is now that the Directors or their designated representatives approve the tender and hence all the assumptions contained in it.

There are potentially two distinct processes which can have equal effect on the financial outcome of the contract. The first is to arrive at a total tender figure and the second is to decide where to put the money in the Bill of Quantities where the contract is not a lump sum.

2.5.1   Derivation of the Tender Figure

The meeting will begin with the Estimator describing the project by taking the Directors through the tender documents. He will then explain the output and productivity assumptions made in the programming and pricing. During this exposition he can expect a barrage of questions.

The Estimator will then present the savings resulting from materials and subcontract prices received after the bills were totalled, which will be automatically deducted from the tender figure. The meeting now moves away from calculation into the field of commercial judgements. All the following considerations will be in the mind of the Director or Manager settling the tender, but obviously they will not all apply to every contract:

1. the Summary of Subcontract prices will be scrutinised to see if there is a reasonable consensus of prices in each discipline. If there is no known reason why a particular subcontract price included in the tender is, say, 20% lower than the rest, it is a risk which might result in an addition;
2. opportunities for deals on crucial items will be explored to reduce the Main Contractor's risk. These are only attractive if the subcontractor is of some substance;
3. a decision will then be made on the potential buying savings on the remaining subcontractors as a lump sum deducted from the tender total;
4. similar decisions to those considered under 2 and 3 are made for materials;
5. the company may be one which considers profit against capital employed and the intention is to front load, which is discussed in the next section. The Estimator will be asked for the cash flow prediction and the Directors may consider discounting some of the prospective earnings on the positive cash balance;
6. the potential overheads and the profit contributions from expenditure of prime cost and provisional sums will be considered and may be included within the overall mark-up for overheads and profits;
7. commercial advantageous savings that are known, such as an exclusive agreement for use of a crucial spoil tip or major specialist

plant, might allow the Directors to increase the tender sum by, say, 50% of the perceived advantage;

8. the meeting will then consider the levels of risk involved in the project and assumptions made. Certain forms of contract carry higher risks. Complicated projects are likely to carry a high rating. For a simple project like a large distribution warehouse with hard standings the Directors might be prepared to take money off at this stage. They would also consider the known reputation and attitude of the Employer and Engineer and the project's location. When deciding the formal tender total the Directors may increase the tender figure if they feel that the Employer or his representative under the contract are likely to be unreasonable or overbearing or if the location may be especially difficult;

9. having made all these adjustments the last figure to consider is the contribution required from the project to the Company's total overhead and profit. This decision is influenced by the Company's financial strength, its forward workload, and the market conditions.

A formal tender figure is thus arrived at. If the contract is a lump sum, whether or not Design and Construct, nothing further is required. (The exception is if there is an Activity or Milestone Schedule attached which gives the ability to apply some front loading to finance the project (see 2.5.2 item 4 below).

2.5.2   Completion of the Form of Tender and Bill of Quantities

The Building Industry, when a JCT form of contract is used, has the advantage of submitting at tender stage only the total figure inserted in a Form of Tender. They are required to submit their Bills of Quantities only if they are one of the lowest two tenders. They therefore have time to make the bills more accurately reflect the decisions taken at the Tender Review meeting.

Main Contractor Tender Reviews are usually held on the day before submission as they must be able to take advantage of the best subcontract prices. Some subcontractors in turn do not want to show their hand until the last possible moment.

Computers do enable broad percentage adjustments to be made rapidly, but they are not quick at selective adjustments. Even if they can be used, most public clients will not yet accept an electronic bid or a print out as the submission and a mammoth manual copying exercise is required against the clock. On a large Bill of Quantities this takes up an inordinate proportion of the restricted time available for tender preparation.

The usual solution to these conflicting constraints is laid out below and considered against Bills of Quantities prepared on CESMM and the Standard Method of Measurement for Road and Bridge Works.

1. General Considerations

The preferred method of distributing the monies is as follows. Firstly to avoid altering the rates already established in the Bill. Part of the contract profit and preliminaries may then be included in the rates because of the savings gained on material and subcontractors. Bill rates are only altered for specific critical late quotes or tender deals (e.g. items like piling where alteration in length could mean paying the subcontractor more than was received.)

The second consideration is to establish in a time-related item a running cost per week which includes a 15% overhead and profit recovery when these rates are used to evaluate any extensions of time awarded under a subsequent contract.

The total of the measured bill and time related-items are deducted from the tender total to arrive at the remaining fixed cost. This is normally put in a front loading item such as erection of offices for the Contractor.

Sometimes, because of the size of the deductions, this last figure can be negative. The opportunity remains to front load to the extent assumed at the tender review (see 2.5.1 item 5) by inserting the desired sum in the erection item and then a corresponding negative figure in the balancing item.

2. Documents Providing for Pricing Methods

The best known of these is CESMM where method-related items and charges are a very flexible and useful tool to both the Engineer and Contractor. The Contractor is required to identify his time-related cost for the evaluation of extensions of time. This is of great value to the Engineer in evaluating potential costs or claims for unforeseen circumstances.

The main advantage to the Contractor is the possibility of an exact match between rates of expenditure and recovery. For example, the method related charges on one contract included an item for purchase of a large quantity of temporary sheet piling and then a credit on its projected sale when the cofferdams were removed.

3. Other Methods of Measurement

Most Methods of Measurement, unless modified, do not specifically identify large slices of the temporary works costs. They are deemed to be included in the rates and the Contractor is not instructed where to include them.

One project for a foot bridge over a sizeable river had cofferdams which resulted in a total prelims figure which was 45% of the contract sum. In preparing the tender these costs had been spread by computer across all the quantities. There was a further problem in that there were a considerable number of provisional quantities which the computer did not identify, thereby potentially depriving the Contractor of the monies needed to execute the temporary works.

4. Activity Schedules or Milestones

Although some contracts with activity schedules or milestone payments also have standard rates to be priced for add and omit variations, in the majority of the cases variations are dealt with at cost. The Contractor therefore again front loads the earlier activities to ensure positive cash flow and the minimum of capital employed. If an activity could be omitted in its entirety the Contractor will avoid any mark-up on that item.

The foregoing demonstrates the need to study the tender papers to discover the actual intent and the time allowances for executing any item. As far as the Client is concerned the contract documentation means that the individual rates inserted are deemed to cover the item inclusion in the respective method of measurement.

Most public clients insist that bids must totally comply with their tender documents or suffer rejection. Some allow qualified bids provided a conforming bid is also submitted, such that they may judge the extent of the saving and any extra risk being passed to them. The form of tender performs three main functions:

1. It lists the drawings and documents comprising the Contract. The Contractor is bound by his tender as soon as he dispatches it to the Client. The Client is not in any way bound to the contract documents until after his letter of acceptance has been received by the Contractor. (This distinction is very important in case there is a major change in market prices after the Contractor has dispatched his tender (e.g. increase in steel prices in 2003-04).
2. It provides inter alia for payment on the basis of rates and the requirement for remeasurement.
3. A contract exists when a letter of acceptance is received by the Contractor.

In the Appendix to the Form of Tender the Contractor is normally expected to fill in his percentage for adjustment of prime cost sums. If it is only presented here and not as a bill item then most Contractors would quote a fairly high percentage as it does not directly affect the tender total. However, this may be a disadvantage if, at tender appraisal, this item is given greater importance.

In a Measured Rate contract the tender total is not such a primary consideration as it used to be when the Contractor could be held to his tender sum.

The notes to the Appendix to the Form of Tender provide for the Contractor to insert a list of materials for vesting, or this may be restricted in the original enquiry document. In the latter case, if there are NIL entries in this section it means that the Contractor will not be paid for materials off site, particularly if, as is usual, the Instructions to Tenderers say that any amendments to the tender documents will disqualify the tender.

CHAPTER 3

Starting Up a Project

3.1   INTRODUCTION

This subject has been taken next to draw together the several management strands which come together on the award of a contract. This is the most critical phase of any project and the time at which the Site Team is really stretched. Contractors normally only obtain one in eight of the contracts for which they bid and it is unlikely that the Site Team has been involved pre-award. The following tasks must be covered by the Site Manager and his team:

- study the drawings, specifications and tender and understand the way the company intends to undertake the project;
- on Design and Construct Projects decide, allocate and brief the Design Team;
- prepare the contract programme on contracts where this is relevant;
- consider the major subcontractors and obtain approvals from the Employer's Representative;
- submit notices for any information to be provided by or on behalf of the Employer;
- establish cost systems and prepare budgets;
- execute temporary works designs and method statements and submit for approval;
- allocate and brief staff;
- mobilise plant, equipment and materials;
- dispatch Form F10 to Health & Safety Executive.

3.2   THE TENDER

Having followed the Estimator's progress through the tendering process the reader will now have had demonstrated that the tender contains all the assumptions, resource levels, planned outputs and methods necessary to construct the project and that these have been approved by the Directors or their representatives.

It is the Site Team's task to make this plan a reality. Therefore, the Site Team must be aware of its provisions at all stages. The Site Team should defend all the assumptions to the Employer's Representative, even if they are marginal. Site

Teams should be required to consult their superiors before undertaking an operation if they consider that it cannot be done in the manner envisaged, or with the resources provided for in the tender.

3.3   ESTABLISH AND BRIEF THE DESIGN TEAM

There are the following items to decide on Design and Construct contracts:

- in the UK the Contractor may also have to appoint the Planning Supervisor with checks on PI and previous experience in accordance with the requirements of the Construction Design and Management
- regulations (see Chapter 11);
- the appointment of a Designer or Designers again in the UK with checks on PI and previous experience in accordance with the requirements of the Construction Design and Management Regulations (see Chapter 11);
- the extent to which subcontractors are to carry design responsibility with the same checks as for Designers;
- establish the Design Brief and the required timing of deliverables;
- establish controls for developing, controlling and monitoring the development of the Design.

3.4   THE CONTRACT PROGRAMME

Most forms of contract contain a Contract Programme. It is the only part that the Contractor is able to write after tender acceptance. As is covered in Chapter 8 and the Contract Appendix, the Contract Programme, together with accompanying Method Statements, demonstrates the Contractor's intent. It is therefore essential that these intentions are mirrored in the Programme.

Increasingly in today's competitive market Contractors price and gain work by intending to complete it in less time than the period allowed. This should always be shown on the Programme with the difference in time being clearly marked, preferably with a separate bar, as "Contractor's Float: No Resources".

It is sometimes argued that, when writing the Programme it is better to absorb the entire period of executing the work. The argument runs that all delays will still be identifiable as separate and distinct periods entitling the Contractor to extensions. The Contractor, it is said, will also avoid being behind programme if he adopts this method. In the real world disruptive delays seldom stop all work in an identifiable area. Their effect, and the entitlement to any extension of time, is often disputed by the Employer's Representative. It is therefore unwise to absorb all available time so that the Contractor can be in a position where he can be ahead or on programme, according to the Contract Programme, but be behind the Tender Intent and losing considerable amounts of money.

The Employer's Representative is bound to point out in any dispute over the effect of a delay that it cannot be significant if the work is finished ahead of or on programme. Where the planning and pricing have provided for continuity of operations for a subcontractor, a particular expert gang, special plant or the use of temporary works materials, it is wise to emphasise this on the programme, by use of demand lines linking the activities on the bar chart.

Where subsidiary areas of work are started earlier than necessary it is wise to show the float before they become critical, indicating where that is with a demand line.

The preparation of the Contract programme should go hand in glove with the resourcing and budget (discussed later) so that the programme submitted has balanced resources with no avoidable peaks. The resourcing itself should not be submitted to the Employer's Representative unless a requirement of the Contract. One must be sure that the true effect of delay to any section of the work is demonstrable from the programme before considering alterations to the actual intention that would gain commercial advantage.

Where appropriate, lead times for information for provisional sums, PC sums, bending schedules, etc., should all be shown on the programme, together with order periods for critical materials on long delivery.

3.5   SUBCONTRACTORS

When preparing the Contract programme it will become obvious from the tender programme the order of precedence of the various subcontracts and which ones must be let immediately. In most cases a new enquiry will be needed, after tender acceptance is received, or considerable negotiation conducted, to ensure that the tender assumptions and conditions are included. The process of letting a subcontract is covered in Chapter 5 and the Subcontract Appendix. Whether or not approval is required under the Main Contract is in the Contract Appendix on page 165.

If there is a major subcontract which, although not required immediately, is central to the entire contract then negotiations should start immediately, before the period for acceptance runs out. Thereafter the subcontractor might attempt to revise adverse prices or conditions in his bid.

It is essential at the time of letting the subcontract to write in directions or get formal recognition of any known restraints on the subcontractor's activities while the commercial advantage is with the Contractor.

3.6   SUBMISSION OF INFORMATION REQUESTS

It is generally accepted that a Contract Programme, which also indicates dates for information, is not sufficient when making a claim for lack of information. It is therefore vital that, at the beginning of the contract, a proper notice be sent listing the detailed information required, together with realistic dates. Lists should also be submitted giving dates for approvals of method statements, drawings and subcontractors.

It is good practice to then update this notice for submission at progress meetings with the Employer's Representative. The likelihood of this being a major issue is only on industrial developments or similar projects where time is the essence of the contract or where the detailed design is developed during the contract. Although public sector infrastructure contracts are normally fully designed, additional information or amended drawings are often required during works progress.

3.7   ESTABLISH THE BUDGET

The Company will almost certainly have its own cost system which the Site Team is required to follow and submit a budget at a date close to commencement.

The Site Team, when taking over the tender to begin construction, must prepare a budget. In an ideal world this would obviously be the same as the tender on Day One of the project. This is seldom the case, however, and the Site Team must recognise that its first budget is his only chance to highlight any variances in his view of the contract from that of the Estimator before his own actions muddy the waters. Hence, in case the company systems do not require budgets at an early stage, the prudent Site Team will do one to ensure that its proposed modifications to the tender plan do produce a better result.

The budget is the Contractor's internal document by which he will monitor performance. It incorporates the tender review decisions and derives rates and items that differ from those in the Bill of Quantities (see Chapter 4).

3.8   SUBMISSION OF TEMPORARY WORKS DESIGNS AND
         METHOD STATEMENTS

In the working up of tender designs and methods to fully fledged schemes there will invariably be some changes and, one hopes, improvements. However, the Site Team should at all times maintain close touch with, and ideally direct, the technical departments of the company to ensure that proposed schemes are not more costly or require more resources than assumed at tender. If the proposed schemes are submitted to the Employer's Representative without qualification they will be deemed to be in accordance with the Contractor's tender intent and with the conditions being designed for when the tender was submitted.

This can effectively shut the Contractor off from reimbursement for changes subsequent to tender submission. In order to mobilise the contract rapidly the Site Team must balance the need for quick submission of the worked up scheme against the equally important requirement to ensure that what is submitted does not adversely vary the assumptions made at tender.

If the fully fledged schemes do vary from the tender designs then the Site Team should follow the procedures outlined in Chapter 9 and the Claims Appendix.

3.9   ALLOCATE AND BRIEF STAFF

More than any other sphere, construction is a people business. Assembly of the correct staff at the correct time is essential to the success of the project. The assembly of the Site Team will normally be an Area or Contract Manager's direct responsibility. The Project Manager will be able to express a view and he should fight to ensure that he has the best possible team around him.

Nobody, at any level, works for money alone and in managing and directing the staff the senior members of the Site Team should recognise the motives and aspirations of the team as a whole. Although larger companies have specialist Personnel Managers, the Project Manager and the heads of the different disciplines are the prime movers in making things happen. If not already part of a company's staff development policy it is beneficial to have a yearly formal review of each staff member's progress and aspirations. This subject is included in this chapter as the Project Manager would be well advised to go through the process with new members of the team to establish their strengths, weaknesses and experience to date.

From personal experience of two separate systems of assessment and review the best results were obtained from the following format. A proforma is issued to the individual with the following sections for him to fill in before the interview:

- the past year's experience and achievements;
- particular strengths and weaknesses;
- perceived training needs;
- desired career pattern for the next five years.

The reviewers will then have spaces to fill in during the interview on recommended off site training, areas on which the individual should concentrate and an agreed development plan for the year. It is the Project Manager's responsibility to see that the company's side of the bargain is kept.

All staff must be briefed on the tender assumptions, if not in financial form, definitely in terms of production data. The number of linear metres of completed drain with a three-man gang and 360 degree excavator and the assumed size of the distribution gang are examples.

Proper planning and management techniques of the type discussed in Chapter 4 are only effective if the senior members of the Site Team do not hoard knowledge. The Site Team must provide a framework for constructive and creative work.

3.10   MOBILISE PLANT AND LABOUR FORCE

The Site Team must ensure that all plant and equipment is delivered to meet the dates shown on the contract programme to enable it to execute the work to time and, if the project is delayed, to be able to claim reimbursement from the Employer. Hire need not necessarily start immediately on delivery. Lack of essential equipment on site, even if there is undisclosed float in the programme, will prove a powerful weapon in the hands of the Employer's Representative when discussing the effect of later events. The types of labour force are discussed in Chapter 6, but on the assumption that some labour is to be employed directly rules must be set for taking on men. These should include:

- a check by management on references;
- a statement of site conditions, rates of pay and incentive schemes;
- that wearing a helmet in designated areas and compliance with all safety rules are conditions of employment;
- that front line management is aware of the right of an employer to remove any man as unsuitable in his first six working days with two hours' notice. Systems which require front line managers to positively affirm that they want the men allocated to them will pay dividends.

The details of necessary management attitudes are covered in Chapter 6 on the Management of Labour.

3.11   DISPATCH FORM F10 TO THE HEALTH & SAFETY EXECUTIVE

In the UK the Form F10 advises the Health & Safety Executive of the nature of the works, gives an approximate number of operatives and the site's location. Its dispatch is a statutory requirement.

In with all the other headings covered above should be a detailed consideration of safety and the working practices to be adopted.

Chapter 11 on the CDM Regulations includes a checklist of information that the Client and Planning Supervisor are required to supply in the UK to permit safe design and construction of the works and at this stage it should be checked for adequacy.

3.12   CONCLUSIONS

The foundations for success, or indeed failure, are laid in the first few weeks from award of contract. Anything left to chance by not fully identifying and delegating the necessary tasks, will, in accordance with Murphy's law, go wrong. The following chapters cover the specific management subjects in detail.

CHAPTER 4

Cost Control

4.1   INTRODUCTION

There are two aspects to cost control, which are:

- control of the resources at all stages of construction; and
- financial control.

As previously set out, to do the first the site team has to have full knowledge of the tender philosophy and productivity which must be allied to proper programming procedures. These matters are detailed in the section on Control of Resources. Overseas main contracts often use all direct resources and the principles are the same for most subcontractors controlling their own direct labour.

The financial control systems are the means of ensuring:

- forecasting of cash flow and contract outcome;
- maximisation of receipts; and
- control of payments.

The importance of cash flow cannot be overemphasised. Contractors in the UK work on small profit margins which, after overheads, can be as little as 2% or less of turnover. They are far more profitable when judged by the returns on the capital employed. The strategy of most contractors is to front load their tenders to ensure that by the second month the contract is cash positive so that the capital employed is minimal.

Standardisation of reporting of the financial outcome with the aim of objectivity is necessary for any company so that the future is clear at any point. It is also essential to enable the company's Senior Managers to take action at a time when they can affect the outcome.

The civil engineering side of construction has followed the building and process arms in to greater division between the control of resources by Site Managers and the financial control by Quantity Surveyors. If divisions build up between the two disciplines on any project it will inevitably adversely affect the financial outcome.

Control of cost is exercised by whoever puts the resources to work. Financial control, other than letting subcontract packages and pricing of employer variation requests, is retrospective and cannot recover monies spent.

4.2   CONTROL OF RESOURCES

These disciplines are to be followed by the construction side of the site management team and impact directly on the outcome of the project.

4.2.1   Programming and Planning

It is generally recognised that there are three levels of programmes; these are listed below:

>   Level 1: The Contract Programme
>   Level 2: Detailed Programme
>   Level 3: Specific Complex Sections

The Level 1 programme is normally the Contract Programme. This will usually show each area trade or activity as a single bar on a Gant Chart. The number of activities will be determined by the size and complexity of the project and any instructions in the Contract regarding what is to be monitored or included on the Contract Programme. There should also be Logic Links showing the physical and resource constraints in constructing the works. (Chapters 3 and 8 deal with the issues that should be addressed in the Contract Programme).

Level 2 programmes should show the order and sequence on which each substantial element of the work is executed. They should be compatible in terms of periods for each area of the work within the overall Level 1 duration.

The Level 2 programme on smaller projects will be drawn up by the Site Team. On larger projects Level 2 programmes may also be contract documents. It is essential in these cases that the productivity assumptions within the tender are still reflected in greater detail. It should show the restrictions on each trade or activity.

On major projects the Level 3 programmes would break down the Level 2 programme into greater detail. It is often not necessary for there to be a Level 3 programme for the whole site. On most sites the place of the Level 3 programme is taken by the weekly programme.

4.2.2   Benefit of Computer Programming

There are many computer software packages allowing programming to be done on computer which can also generate progress reports.

Although it takes longer initially to produce a programme with full logic links and resources, it leads to considerable benefits thereafter.

> 1.   Resource Levelling
>   The more advanced software packages permit resource levelling to be undertaken, altering the activity durations with obvious benefits for efficiency.

2. Package or Trade Programme
   Using filters the software can produce package or trade programmes from both Level 1 and Level 2 which can be incorporated into subcontracts and be used to monitor their progress.
3. Ease of Review
   This is in terms of predicting the effect of current events and of delay to the project and also of evaluating employer change requests, all of which is discussed below.

It is practically impossible to run a substantial project under the NEC form of contract without access to computer programme software, indeed the software to be used is often specified in the Works Information.

4.2.3   Managing Change with Revised Programmes

Software packages can permit two important functions to be carried out easily:

1. To draw a 'time now' line on the programme and automatically move all incomplete activities in the same logic. This shows what must be done to complete a project within the original or extended time or predicts a new completion date with some accuracy.
2. Enables the effects of variations to be either established or predicted. This is either retrospective from the item above or can be done by lengthening the duration of the affected element on the programme.

4.2.4   Setting Direct Resources to Work with Relevant Production Targets

In many instances, the resources will be put to work by means of weekly programmes or Level 3 programmes. The productivity assumed in these should be equivalent to the tender.

Where work is being executed with directly employed resources either the Contractor's or Subcontractor's front-line managers must be provided with productivity information that they have to achieve. This can be in the form of number of units, square metres, cubic metres or a particular stage in the operation with a set amount of resource and should be available for each operation.

The service gangs and resourcing should be specified in terms of numbers and durations. As set out in Chapter 2 at 2.4, The Pricing Operation, the assumed productivity is inherent in building up the rates and should be easily available.

The staff benefit because they know when they are succeeding and the project benefits because corrective action can be taken on the next day, provided it is part of their remit to report if the targets are not achieved.

Identification of the changed circumstance, if the lack of achievement of the target is not the Contractor's or Subcontractor's liability, leads to early identification and notification under the Contract or Subcontract and correct allocation of resources in the records (see 4.2.6 below).

Production staff can be content that they are succeeding if they achieve productivity targets and remain on programme or identify and record on allocation sheets the causes and resources involved.

4.2.5   Directing Subcontractors

This is covered in detail in Chapter 5. Front line managers must be aware of the agreed attendances and facilities to be provided and ensure that they are in place before the Subcontractor mobilises to site.

Notoriously, front line managers sign hours record sheets for extras which are truly included within the package. Systems need to be in place to ensure that nothing is signed unless the cause is known and recognised by the Project Manager as being a true extra.

Front line managers must be made aware that nothing can be deducted or omitted from a Subcontractor's package without proper notice.

4.2.6   Resource Records

This section covers records which are required on every project. Specific records required for financial claims or evaluation of variations are covered elsewhere.

Plant Returns

The main purposes are to calculate the expected cost (to accrue) of plant ahead of an invoice and to provide a check when invoices arrive that they are for the correct period and equipment. The Conditions of Contract often require submission of Plant Returns.

Goods Received Sheets

Similarly a log of all material delivery tickets to accrue for the expected cost and to ensure that data is available for invoices to be checked.

Labour Returns and Allocation Sheets

The Labour Return is also normally a Conditions of Contract requirement but also provides the basis for payment of labour. The most important aspect is the correct and detailed completion of Allocation Sheets showing what the labour and associated resource has been deployed on.

Wherever there are variations it is essential that these are identified and properly recognised.

4.3   FINANCIAL CONTROL

The purpose of this section is to provide an understanding for production managers of the disciplines and processes necessary for full financial control and certainty of outcome. It cannot, in the space allotted, be a comprehensive manual for Quantity Surveyors performing the duty.

4.3.1   Forecasting of Cash Flow and Contract Outcome

Major contractors have now mostly adopted electronic reporting with linked spread sheets building up to a summary sheet which is reviewed monthly by the Contractor's directors. These are termed in most companies as Cost Value Reconciliations or CVRs. A typical summary sheet is set out below and is now explained, starting with the less obvious cost categories.

DOM SC   Domestic Subcontractors who carry the full contractual liabilities that the Contractor himself takes under the Main Contract.

NOM SC   Nominated Subcontractors where the Contractor has less risk.

DESIGN   Subcontracts where design responsibility is passed to the Subcontractor.

LOSC   Labour Only Subcontractors who are paid for labour supply supervised by the Contractor.

L&P SC   Labour and Plant Subcontractors who are paid on rates to execute items like concreting or formwork who do not supply materials. They have liability for quality and programme.

PRELIMS   Preliminary items include staff, site set-up and dismantling. Service labour and plant are sometimes included in this category.

MARGIN   In here would go the overheads and profit on employer-designed contracts but the example below, for a Design and Construct contract, assumes that this will be made out of buying gains and savings on the contingency.

FEES   Normally for design.

CONTINGENCY   The size of contingency in the example below recognises a Design and Build contract where the Contractor includes an allowance for any shortfall in his tender assumptions. On employer-designed projects this would be the Provisional Sums.

FORECAST

	BUDGET	ADJ	BUDGET		PROJ	MARGIN
	TENDER		AFTER	EXP	EXP	
	ANALYSIS		ADJ's			
LABOUR	1000000	-900000	100000	4000	95000	5000
PLANT	50000	-40000	10000	2000	9500	500
MATERIALS	450000	-350000	100000	5000	95000	5000
DOM SC	1000000	750000	1750000	5000	1662500	87500
DESIGN NOM SC	500000	450000	950000	0	902500	47500
LOSC	0	25000	25000	0	23750	1250
L&P SC	0	75000	75000	0	71250	3750
PRELIMS	500000	-10000	490000	30000	465500	24500
MARGIN	0	0	0		0	
FEES	250000	0	250000	15000	237500	12500
CONTINGENCY	250000	0	250000	0	237500	12500
TOTALS	4000000	0	4000000	61000	3800000	200000

4.3.1.1   Example at Project Commencement

The summary sheet shows Week 4 of a 52 week contract and for simplicity this is assumed to go through without variations to Week 52 to demonstrate the process.

The second column is Budget: Tender Analysis which demonstrates the expected expenditure and recovery.

The third column is headed ADJ for adjustments. The reasons for these adjustments are to recognise the alterations that can happen to the pricing after Contract. This is more prevalent in building than civil engineering.

It is common for the Contractor to price many of the basic operations such as ground works and frame construction at rates from his data bank and subsequently seek subcontract prices on award of the contract. The negative adjustments in Column 3 represent a shift of all production activities to Domestic, Labour Only and Labour and Plant subcontractors. The reduction in prelims is the Domestic Subcontractor taking on setting out.

These adjustments are applied to Column 2 and produce a revised budget in Column 4. It should be noticed that the theoretical value is £4 million for both the tender and the budget.

Column 5 contains the actual expenditure which will be taken from the Contractor's cost reports. These can be inaccurate and depend upon the accuracy of both coding and site records of resources (see Cost Appendix on Debugging the Costs).

Column 6 the projected expenditure which leads to the generated margin. Each cost heading will have a linked summary sheet of its own and beneath this will be further linked spread sheets for each element of the work. These will start from the current cost system information in Column 5 and it is here that any unexpected good or bad news should be reserved out until it is explained. Alterations to the base information will automatically feed through to the summary sheet which will be reviewed by the Directors once a month.

4.3.1.2 Example at Project Completion

The table below shows the final situation on completion in Week 52 with the Expenditure and Projected Expenditure as the same figure. A typical scenario is shown with increased directly employed Labour, Plant, Materials and Preliminaries offset by reductions in Subcontractor payments and savings on the Contingency.

The unlikely situation of the same contract figure is shown with the margin being generated solely by improvements on the cost of executing the works. "Management of Change" below will detail how this would be dealt with.

FORECAST   MTH YEAR >> WEEK NO 52 >>>   Report date

	BUDGET	BUDGET	BUDGET		PROJ	MARGIN
	TENDER	ADJ's	AFTER	EXP	EXP	
	ANALYSIS		ADJ'S			
LABOUR	1000000	-900000	100000	99000	99000	1000
PLANT	50000	-40000	10000	12000	12000	-2000
MATERIALS	450000	-350000	100000	120000	120000	-20000
DOM SC	1000000	750000	1750000	1500000	1500000	250000
DESIGN NOM SC	500000	450000	950000	850000	850000	100000
LOSC	0	25000	25000	25000	25000	0
L&P SC	0	75000	75000	75000	75000	0
PRELIMS	500000	-10000	490000	500000	500000	-10000
MARGIN	0	0	0	0	0	
FEES	250000	0	250000	237500	237500	12500
CONTINGENCY	250000	0	250000	181500	181500	68500
Totals £	4000000	0	4000000	3600000	3600000	400000

4.3.2 Management of Change in Forecasting

4.3.2.1 Contractors Own Fault Delay

For accurate reporting of potential costs most companies require their Quantity Surveyors insert for every week behind programme an additional week's preliminaries, labour, plant and LOSC cost into the Projected Cost. This would only be removed when a new programme is produced by the site which shows the measures to be taken to complete to time. Any costs of these measures would then replace the initial estimate. It is failure to do this that leads to nasty surprises and combined with the treatment of unagreed items (see below) is the biggest cause of the ultimate failure of contractors.

4.3.2.2   Costs Associated with Ordered Variations

Several of the Conditions of Contract reviewed in this book require that the Contractor give estimates before variations are instructed and this is indeed good practice. The Quantity Surveyor effectively acts as estimator; where there are Bill of Quantities he may be required to use analogous rates. In other instances he will build the price up from first principles, as described in Chapter 2 "The Tender". An essential element in the pricing is to check whether it has an impact on completion. This can be done (as is required in ECC contracts) by inserting the appropriate duration into the computer based logic linked programme and seeing whether end or sectional dates are altered.

Where they are prelims these are also added to the quotation. If accepted and implemented, the relevant elements of the budget are increased, as well as the increases to the projected expenditure.

4.3.2.3   Unagreed Variations and Claims

The treatment of these items is where financial reporting can become unreliable. At one end of the scale there can be extreme optimism where it is assumed that all sums will be paid by the Employer, and at the other overprovision.

Take, for instance, the subcontract reconciliation. The Subcontractor would receive the figures listed as certified amounts:

Subcontractor's Name: *Partitions Ltd*

Contract Sum	Order	Certified
Partitions	10,000.00	*9,000.00*
Dry lining	25,000.00	*24,000.00*
Suspended ceilings	5,000.00	-
	40,000.00	

Agreed variations

1. Additional s/c to Room 4	800.00	*800.00*
2. Additional s/c to Room 8	200.00	*200.00*
	1,500.00	

Certified   *34,000.00*

Overview

Delay and Disruption Claim   6,000.00

Contingencies
1. Non claimed wasted
   materials   200.00
2. Contractor attendances
   performed by Subcontractor   500.00

Gross Liability   48,200.00

Obviously, the agreed variations attach to both the value (or budget) and the projected expenditure.

The first decision is whether to take all of the Disruption Claim into projected cost as has been done in this case. On the same issue, the amount that is put against value can vary between zero and £6K. Contingencies are unclaimed items that, in this case, have been taken into projected expenditure reducing the profit margin.

A considerable difference in reporting can be experienced for the same situation from different individuals unless there are consistent guide-lines across the Contractor's staff when similar items occur in most subcontracts.

The Contractor's staff also have to assess the sum that is expected to be recovered from its own claims for prelims, etc. Managers under pressure sometimes are optimistic to avoid immediate internal criticism and do not make proper provisions. The direct result is that the Directors are not alerted at a time when they can have some input into correcting the situation.

4.3.3   Maximisation of Receipts

This involves three disciplines, being:

- Early submission of fully priced and supported variations;
- Interim claim submissions to stimulate on account payments; and
- Submission to the required dates and chasing of certified payments.

The first two require full knowledge of the contract and its notice procedures as well as the tender assumptions. Only proper co-operation between Production Managers and Quantity Surveyors will lead to the best financial outcome.

4.3.4   Control of Payments

The control of payments begins with the subcontracts and orders to ensure that the Subcontractor submission dates correspond with the Main Contract dates and that payment terms permit the money to be received from the Employer before the Subcontractors and Suppliers are due payment.

Material orders are normally 60 days and should be rigorously enforced.

Checking systems should require that invoice queries be sent out whenever these do not match Plant Returns, Labour Records or Goods Received sheets.

CHAPTER 5

The Management of Subcontracts

5.1   INTRODUCTION

Successful management of subcontract work begins well before any work starts on site with the choice of the packages to be subcontracted and with care in the selection of subcontractors.

A checklist for subcontractor selection is included in the Subcontract Appendix. Whether or not approval is required under the Main Contract is in the Contract Appendix on page 165.

Once potential subcontractors have been chosen the next step is to produce full, detailed and clear enquiries (a checklist for subcontract enquires is included in the Subcontract Appendix). These are an essential precursor to a well constructed subcontract document which leaves neither party in any doubt as to the work to be done, the time it is to be done in and the physical conditions under which it is to be done.

To achieve this precision the drafting of the subcontract must be executed with knowledge of the methods, sequences and programming that are actually going to be used on site. A pre-let meeting is the normal way that the conflicting elements of the enquiry and the Subcontractor's quote are resolved.

In Chapter 3 on Starting Up a Project it has been explained why changes will almost certainly be required from the tender enquiries because these were issued of necessity before project planning was complete.

Once the Subcontract has been let then Site Production Management must be made aware of the attendances and the budget, adjusted as particularised in Chapter 4.

I have tackled the subject by starting with Nominated Subcontractors, where the Contractor might not enter into a subcontract at all, and then proceeded to letting a subcontract.

I have included the detailed analysis of the CECA Blue Form of Subcontract and commentary on the NEC Subcontract Form in the Subcontract Appendix. This leaves within this chapter the principles behind the subcontract and the contractual action required of both the Contractor's and Subcontractor's Project Managers.

The management of major subcontracts by using the central tool of properly constructed and minuted progress meetings and the realities of Labour Only Subcontractors are taken last.

5.2   NOMINATED SUBCONTRACTORS

Nominated subcontractors are used by Employers and their representatives to obtain either one or all of the following advantages on projects where the Employer executes the design and the form of contract provides for this eventuality (see Contract Appendix):

1. Detailed design advice on specialist items which will affect the design of the works to be carried out by the Main Contractor.
2. To obtain particular plant or equipment which matches an Employer's existing plant.
3. Detailed control of cost.
4. Early start of items with off site fabrication and long lead in times which would extend the time required for the project.

The clauses relating to Nominated Subcontracts are contained within the Contract Appendix where it is detailed why the Contractor has a right on ICE based contracts to object to any nominated subcontractor who broadly cannot or will not:

1. Meet the main contract specification or Contractor's programme.
2. Accept the full provisions of the main contract particularly with regard to damages.
3. Show proper and adequate insurances.
4. Accept the forfeiture provisions of Clause 63 of the main contract.
5. Meet the criteria required by the CDM Regulations (see Chapter 11).

Most cases of objection arise from the programme and the conditions of the main contract and the Project Team must ensure that the Subcontractor is totally back to back with him. This is done by using either the CECA or another recognised standard form of subcontract. Refusal to accept a standard unaltered form of subcontract would normally be valid grounds for objection to "the nomination". The Contractor could similarly not insist on anything more onerous. In the broader wording of condition 1. in the contract document it is also possible to object to a Subcontractor whose finances are suspect.

5.3   LETTING A SUBCONTRACT

Having prepared the contract programme and a fully resourced programme for budgeting purposes, the Project Manager will now be fully aware of the package of work he wants to subcontract and all restraints on the operations. This may well be different to that envisaged at tender and sent to a wider selection of subcontractors (see Subcontract Appendix for criteria for commercial choice and Chapter 11 for details of competences required by the CDM Regulations).

There are several key contractual administrative requirements of the Project Managers of both the Contractor and Subcontractor which are central to the financial outcome for both parties. Because of their importance I have taken these here and leave it to the reader to go through at leisure the detailed analysis of the CECA and NEC subcontract in the Subcontract Appendix.

5.4.1 Principal Contractual Administrative Actions Required of the Main Contractor's Project Manager

1. To instruct the Subcontractor in writing to start on a particular date and, if the time for completion is not stated in the Third Schedule, to give a time by which the works or sections of the works must be complete.
2. All instructions to the Subcontractor must be in writing. The subcontract forbids the Subcontractor to alter the subcontract works without written instruction from the Contractor. Oral instructions from the Engineer must be confirmed to the Subcontractor in writing as well as to the Engineer.
3. Carefully monitor the Subcontractor's quality and adherence to programme and inform him in writing of any breaches.
4. Notify the Subcontractor in writing of all extensions of time granted on the Main Contract which affect or are applicable to the subcontract.
5. In the case of damage to any of the subcontract works prior to the issue of a completion certificate relevant to that section of the subcontract works, write to the Subcontractor requiring him to repair the damage at his own cost, whether he was on site at the time or not.
6. If the Subcontractor does not comply with written instructions, write again giving a time by which compliance must be achieved "or action will be taken under Clause 17 of the CECA subcontract". Advice should be sought (and indeed reference made to the notes in the Subcontract Appendix on clause 17 of the subcontract) before action is actually taken.
7. Within the UK, where Statutory Adjudication applies, any sum set off against a payment must be fully detailed (all contractual bases and separate sums accruing) and submitted or resubmitted to the Subcontractor after the receipt of the particular application and before the due date for payment.

5.4.2 Principal Contractual Administrative Actions Required of the Subcontractor's Project Manager

1. As with the Main Contract the Subcontractor must obtain written authority from the Contractor, who in turn must obtain written

authority from the Engineer, before he sublets any parts of the subcontract works.

2. The Subcontractor's Project Manager must ensure that all instructions from the Contractor are in writing and only instructions from this source are complied with. This is particularly important in cases where the Engineer has obtained tenders for specialist equipment on the basis that they will provide for the design of structures and that actual tender acceptance will be the responsibility of the Main Contractor. During design there may be considerable direct communication between Engineer and a specialist subcontractor. After main contract and subcontract are let this direct contact must cease and all such communication must be through the Main Contractor.

3. The Subcontractor's Project Manager must make all notices and returns required under the Main Contract. This basically means notice of claims and submissions of contemporary records. The potential penalties for non-compliance are high and can include all the costs the Contractor might have recovered if the Subcontractor had complied (Clause 10(3) of the CECA Form but similar provisions in other forms).

4. On the same subject as item 3, if the circumstance is likely to lead to delay it is a "condition precedent" for its recognition that the Subcontractor notify the Contractor within two weeks of its first occurring (Clause 6(2)(c) of the CECA Form).

5. The Subcontractor's Project Manager must ensure that the interim valuations are submitted seven days before the specified date for them to become "valid statements" and eligible for interim payment.

6. If a dispute arises the Subcontractor is required, as soon as it is clear that the matter cannot be resolved, to send a "Notice of Dispute".

5.5   MANAGEMENT OF MAJOR SUBCONTRACTS

5.5.1   Introduction

On major complex projects the subcontract packages themselves may be in the multi-million pound range. These can only be properly managed and controlled by the Project Manager understanding the rudiments of the technologies being used (if not a building or civil element) and being aware of the relevance of all the sequences and methods being employed.

The Main Contractor's Project Manager must never be just a post-box between the Engineer or Contract Administrator and Subcontractor. The most valuable management tool in controlling major subcontracts is to establish regular progress meetings a few days ahead of the Main Contract progress meeting.

A suggested format is outlined below. I have assumed that the Contractor has requested, and received, a detailed programme for both off site manufacture and on site works which has marked on it, or appended to it, all necessary information or approval dates.

Meetings of this type are shortened considerably if the Subcontractor produces a written report covering all the subjects below, which can be commented on where relevant, rather than have every detail reported orally. It should be remembered that a written report can be construed as a notice and the Contractor should challenge anything with which he disagrees.

5.5.2   The Meeting

5.5.2.1   Safety

A separate initial meeting introducing Safety Advisors and passing over all necessary and relevant information and site rules and proceedings can be followed by a review of the monitoring of the achievements on site against the targets set.

The submission of Risk Assessments for any work about to start and compliance with the CDM Regulations will form a major part of this element (see Chapters 11 and 12).

The main purpose of this submeeting is to ensure that the Health and Safety Plan is up-to-date and relevant and that any revisions are issued to all the parties promptly. The results of this element will normally be the first item on the agenda for the progress meeting.

5.5.2.2   Contractor's facilities

Taking any shared welfare first together with a review of submitted risk assessments and results from safety audits allows professional Health and Safety Advisors from either party or even the Client to withdraw as the provisions of the Health and Safety at Work Act have now been covered.

The other facilities and constructional plant are next to be reviewed and a brief statement made that they were adequate. If there is any doubt as to their adequacy, written instructions must be given to the Subcontractor for positive arrangements so that any dispute does not lead to delay, or give rise to the opportunity for hindsight to identify the dispute as a source of delay.

The Contractor should ensure that he acts so that the Subcontractor is able to agree to a minute such as "the provision of craneage for the pump installation is being undertaken by the Subcontractor to meet his requirements and is not a cause of delay, although the Subcontractor reserves the right to claim his costs for this item."

5.5.2.3   Information

The flow of information should be carefully monitored each month with the Subcontractor being required to give dates by which he requires information and by which he will give other information, as it is unlikely that all detailed items will match the original global dates for information.

The situation will be compounded by part answers and further queries which will make this necessary. The Project Manager should write, before each Main Contract progress meeting, to the Engineer with these dates in a formal Notice, thereby ensuring its timely production or protection under the Main Contract.

At the end of the Information Section of the meeting the Project Manager should minute "that other than the matters raised above the Subcontractor does not require any further information".

5.5.2.4   Off Site Manufacture

Where manufacture occurs off site prior to assembly or construction on site, probably by a series of subcontracts to the Subcontractor, a section on the progress of each item should be included. At the end of the section the Project Manager should minute "that there is no area of procurement or off site manufacture which will affect the completion of the site works in accordance with Programme No. XYZ." If this is not possible the offending item should be excepted from that statement and then proposals should be sought for the elimination of the problem or instructions given to overcome it, as is appropriate.

5.5.2.5   Progress on Site

During the course of this section it is important for the record to establish exactly what the position is relative to the programme that the Subcontractor will have been asked to produce to match the period for completion in the Subcontract.

The review of the programme should result in an agreed marking up by each party of the progress against each bar line. The Subcontractor should be invited to state in conclusion his overall assessment of his position relative to the programme. If he is behind, the Project Manager must require proposals for returning to programme or give instructions (which he should follow up in a letter as soon as possible as minutes of meetings are not classed as written instructions).

Both sides should be aware that progress reports submitted at a meeting can be held to be notices and the Contractor must ensure that he agrees with or corrects any statements in the report or if necessary acts upon them.

A statement at the end of this section should be made to the effect that "the Subcontractor is not aware of any item that will cause delay to the subcontract works."

5.5.2.6   Contractual Issues

It is as well to include a forum for airing disputes and noting whether further submissions are required. It also allows the Contractor's team to report on progress in any joint submissions made to the Engineer. These headings form a checklist which can be scaled down when dealing with smaller subcontracts.

5.5.3   Summary

The secret to successful subcontracting of parts of the main works is always to understand the needs of the Subcontractor and to rigorously monitor and apply corrective instructions on quality and progress at an early stage.

Although the subcontract form provides for the Contractor to recover all his costs from a Subcontractor if he defaults, this will not help the Contractor's reputation. Even this provision is not in practice guaranteed because many Subcontractors are not financially strong enough to actually pay the resultant costs of default. The Contractor may be using them in the first place because of their low price and overheads and more than likely small assets. There is no room for laissez faire in the control of Subcontractors.

5.6   LABOUR ONLY SUBCONTRACTORS

These fall into two categories as far as the Project Manager is concerned. The first type provides a set number of men by the hour who are supervised by the Contractor's foreman. Secondly, there is the large subcontract with plant and labour rates against the Bills of Quantities items and descriptions and subject to the Blue Form of Subcontract with all that it entails.

The first category of Subcontractor requires no contractual handling. The resources supplied are effectively direct resources and cost control is exercised as detailed in Section 4.2.4.

The second is dealt with on a carrot and stick approach. Most subcontractors are survivors, who have learnt to protect themselves against the provisions of the subcontract, or unscrupulous Main Contractors, depending on your point of view, by limiting their liability.

Few will accept CECA or other standard payment terms as they will have paid out seven weeks wages and plant costs before the first cheque comes. By asking for fortnightly payments their practical liability is only what they leave in the Contractor's pocket if they walk off site. The threat of this, which both sides appreciate will disrupt the Contractor considerably, causing him costs he will have the greatest difficulty in actually extracting, is the counter balance to the legal armoury of the Contractor.

The circumstances leading to the Contractor choosing Labour Only sub-contracts rather than Direct Labour are covered in the next chapter.

CHAPTER 6

Management of Labour

6.1 INTRODUCTION

There are always changes in the composition of and employment of the labour force in the industry which are driven by employment laws tax law and the industrial relations climate.

In the UK there is the beginning of a move back towards directly employed labour for bulk standard operations by Contractors who have a spread of continuous operations across particular regions.

This will mean that there will potentially be a return of the problems and challenges that occurred in the 1970s and early 1980s when the last general period of direct labour employment occurred.

In the meantime market forces have lead to smaller subcontractors employing small teams or reliance on Labour supply. Bonus and pay rates have been driven by demand for labour and tailored to that necessary to keep individuals. This may well change.

6.2 THE CHOICE AND COMPOSITION OF A LABOUR FORCE

Having decided the major elements that are to be subcontracted, the Project Manager then has to consider how he is to set about employing the men necessary to carry out the work. From his resourced programme he should be able to derive a histogram by trades of the men required.

Certain situations will still lead directly to the choice of Labour Only Subcontractors:

1. Short duration employment for skilled men.
2. Small numbers of men required, not justifying setting up incentive schemes.
3. A known militant area.

A regular practice is to use Labour Only Subcontractors tied to rates for formwork and steel fixing and to employ general operatives directly, who are usually more amenable, and for whom continuity of work is easier to find.

The recognised level of directly employed operatives above which the Project Manager must expect to spend time on formal industrial relations practices is between 50 and 70.

However, by using Labour Only Subcontractors, the Contractor is adding a premium to his labour costs and he should recognise this when considering projects which involve hundreds of men. Over a longer period with continuous work available direct employment must be cheaper given proper control.

6.3   THE WORKING RULE AGREEMENT

Most companies within the Construction Industry employ labour on the terms and conditions laid out in the Working Rule Agreement, which provides a basis for the conduct of industrial relations.

Project Managers often work for many years on sites with small labour forces or Labour Only Subcontracts and are not equipped for the rigours of dealing with a large direct labour force. The first essential is to understand the Working Rule Agreement thoroughly as it will be expertly quoted by the men to the Project Manager.

There are three areas which the Project Manager must understand. Firstly, it provides that until the procedure has been exhausted, there shall be no stoppage of work either of a partial or general character including a go-slow, a strike, a lock-out, or any other kind of restriction in output or departure from normal working. Indeed in the UK an official ballot must be held for a strike to be legal.

Every effort must be made to ensure that these requirements are fulfilled or the difficulties will multiply.

Secondly, the Disciplinary Procedure where written confirmation should be given of verbal warnings, which should be given in the site offices and recorded in the man's record.

The follow-up of written warnings is an essential requirement and must be used. Wherever possible the system should be impersonal and automatic, so that accusations of personality clashes and bias are reduced to a minimum.

It should be borne in mind that, in the UK, if a man is sacked and the procedure has not been followed, even in cases of gross misconduct, the Company is almost certain to lose if the dispute is referred to an Industrial Tribunal.

It is also an important point that the disciplinary procedure can be implemented against persons who do not comply with the grievance procedure.

The third aspect to be fully understood is the operation of the guaranteed minimum, and its possible effect during periods of extreme bad weather and industrial action. This means that if a man works on Monday of week one and is available for work thereafter, but not put to work, he is still entitled to the guaranteed minimum for weeks one and two.

Hence, if the weather is very bad in mid winter it is advisable not to put the men to work on Monday morning for a few hours. Rule X A(2) (d) provides for the suspension of the guaranteed minimum in cases of industrial action in contravention of the grievance procedure and this should be used as discussed later.

The details of the Working Rule Agreement, its explanations, standard letters for all occasions, typical offences and categories for disciplinary procedures are all covered in the booklet issued to member companies of the CECA entitled "Industrial Relations in Civil Engineering Construction".

6.4   LARGE DIRECTLY EMPLOYED LABOUR FORCES

Before considering details there are broad principles which should be adhered to. Firm but fair management is always essential. The advantages accruing to the Contractor through the disciplinary and grievance procedures must be used to the full.

Negotiations must be open and explicit. Agreement by the labour force is contingent on the spirit of the agreement being fully implemented. It is definitely counterproductive to expect that a labour force should stick to the letter of any agreement, as must a Subcontractor, if this is against the spirit of the agreement. The result will only be further disruptive action and a loss of confidence in management.

Given implementation of consultation systems and the direct access of Stewards to the Project Manager, care must be taken to ensure that front line management is at least as well briefed as the workforce. This can be done through the weekly planning meetings, but must remain a prime consideration.

Large labour forces must be subject to rigorous scrutiny during recruitment and the measures described under Chapter 3 (Starting Up a Project) fully implemented. Thereafter the Project Manager should welcome the election of Stewards but should exercise his right to restrict the numbers and object to unsuitable candidates.

Nearly all disputes will arise from one of three sources:

1.   Discipline
2.   Bonus payments
3.   Redundancy.

Turning first to discipline, it is necessary for the Project Manager to ensure that front line managers know and operate the disciplinary procedure. The rules of the site must be clear and enforced all the time. If they are only exercised now and again, in unannounced purges, this will cause resentment and lead to action, whereas the continued imposition would be readily accepted.

Discipline will follow easily on a well run site where the work is going well and materials are always to hand when required by the operative as a result of good planning and its implementation. This will lead to the opportunity to regularly earn bonuses through the incentive scheme and pride will develop in the entire workforce as a result of success. Discipline is difficult to enforce in isolation on badly run sites as the causes of indiscipline are founded in management's weaknesses.

Communication with the workforce is best arranged on a regular monthly basis so that information can flow in an atmosphere that is not affected by dispute or tension. A way of dealing with this is to form Joint Consultative Committees having management representatives (preferably including a front line manager), a full time Union Official and appointed Stewards. The subjects covered can include matters arising from the safety meeting, a review of the progress on the contract and coming events, a review of the Incentive Scheme and prospects for employment.

Bonus schemes are the single biggest cause of disputes on major sites. They are covered later in this chapter, but where it is known that there is going to be a large directly employed labour force on the contract the Project Manager must design or use a proper system from the outset. Correct targets are not enough as the systems can be manipulated to even double the yield for any set level of production. The scheme must be such that gradual erosion of the productivity to reward ratio does not take place.

As the provision of the resources is vital to the earnings of the men it can be useful to appoint hourly paid Trades Foremen who are given bonus payments linked to the trades earnings. They report to the Staff Section Foreman but their main task is to keep the gangs supplied with the "nuts and bolts".

The inevitable occurrence of redundancy on every major project must be carefully managed. Where considered appropriate, termination bonuses may be used and typical schemes are discussed in the section on bonus schemes. The first priority of the Project Manager is to decide well in advance the numbers to be made redundant and the timing of each phase. He should then check on the statutory notice to the Department of Employment and the Unions. If not correctly given the Company could find itself paying for several weeks of guaranteed minimum to those made redundant. It is usual that a labour force will insist on a first in, last out, basis for redundancy within categories. There is little that the Project Manager can or need do about the trades as the skills are similar. He can and must ensure that he has a balanced labour force in the general operative class.

There are many classifications of general operatives with differing plus rates. Before redundancy becomes an issue the Project Manager should ensure that the allocation of men into the plus rate groups accurately reflects the work they are executing. Operatives' skills develop and they may well have changed function since employment. Obviously if all the concrete finishers and pipelayers are made redundant whilst work of this type remains, there is unnecessary disruption.

Alterations of classifications should be done well in advance so that the legitimate objective of a balanced labour force is achieved without disruptive accusations of favouritism and wrongful selection for redundancy, which could form the basis of a tribunal action.

6.5   MEDIUM SITES

The Project Manager should be aware of all the aspects mentioned above but need not force their formal recognition or implementation. Bonus scheme operations are simpler and need not be as elaborate as those discussed below. The targets will be allowed to act unchallenged in isolation or on relatively small portions of the total activities undertaken.

6.6   BONUS SCHEMES

6.6.1   Introduction

There are three main systems, one based on hours saved, another on cash prices for units of construction and the third on achievement of a programme. The straight cash price method when employed by a Labour Only Subcontractor may be the only remuneration paid to the men. With Main Contractors operating the Working Rule Agreement on large sites, the price is in addition to hourly wages. This system, although popular with the men, has the disadvantage that there is no obvious point at which the operative, or groups of operatives, are ceasing to pay their way.

During disputes and periods affected by impending redundancy, reasonable wages can be had for lower productivity. Management is not able to point to a specific and undeniable point at which the operatives are not earning their pay and at which the disciplinary code may be enforced with impunity as outlined below.

Some cash-based schemes do set the number or hours spent on the units at the operatives' hourly rate against the cash price total. This is then essentially the same as an Hours Saved scheme, as management can put the disciplinary code into action as soon as negative bonuses are produced from the scheme.

The Hours Saved schemes are discussed below. The programme system is best reserved for termination bonuses as otherwise there will need to be frequent negotiations as new programmes are produced. These are, in effect, complete scheme negotiations and are bound to lead to disruptions and gradual erosion of the productivity/reward ratio. Termination schemes are discussed later.

6.6.2   Setting a Target

Again the tender will form the best guide as the Company has committed itself to the productivity it expects to achieve. Targets are best not considered in isolation but a group should be produced and any smoothing of rates made before issue of the figures.

Consider the case of fixing and striking formwork using a scheme with a saving rate of £10.00 per hour (see Labour Appendix for details) with the basic hourly rate for trades of £12.00, minimum £2.00 bonus and a further £3.50 for employment costs and travelling, giving a total of £17.50. If we take an item where 1.5 man hours are allowed for striking a particular type of formwork the total in the nett Bill of Quantities (i.e. before "spread monies" overheads and profit) for labour will be shown as

$$1.5 \times 17.5 = £26.25$$

Hence, a target of the actual allowed man hours of 1.5 will lead to the following examples:

If executed in

(a) 0.75 hours Unit cost   = Bonus + Basic Rate + Employment costs
                            = (1.5 - 0.75) 10.00 + 0.75 (12.00 + 3.50)
                            = 7.5 + 11.63
                            = £19.13

(b) 1 hour Unit cost   = (1.5 - 1.0) 10.00 + 1.0 (12.00 + 3.50)
                            = 5 + 15.50
                            = £20.50

(c) 1.5 hours Unit cost   = (1.5- 1.5) 10.00 + 1.5 (12.00 + 3.50)
                            = £23.25

Using this system, as long as positive bonuses are being earned, unit labour costs are bettered and improved with higher productivity (always assuming control in the systems). At the point at which the men earn zero bonus the unit costs are still on the right side. At that point it is easy to maintain that if the men do the work in more hours than the base target, and in effect a negative bonus is produced by the incentive scheme, they are not earning their base wages and the disciplinary code can be activated as discussed later in this chapter.

In setting targets, the Project Manager should consider the whole and possibly reduce the tender allowance to cover the daywork and any non-productive element. Before issuing targets the full range of any type of work should be looked at. If, from converting the tender assumptions, the resulting target is obviously too low there is no point in expecting operatives to accept the company's errors. This should be brought into line before issue by trimming all other targets or any particularly generous ones. Similarly, any generous areas should be reduced to reasonable levels to maximise the company's return. This process will not be possible after the targets have been issued.

The targets, or a set of targets, issued should be for the same items that are identified in the site costing system, if at all possible. This allows continual monitoring of the effectiveness of the scheme.

It is prudent to use a 100% saving scheme for the ease of arriving at the cut off point of earnings to production ratio. The bonus calculator rate must not be exactly the same as the base rate. If this is the case there will be pressure for the yearly increase in base rate to be passed on to the bonus and that will not be recovered under any of the VOP formulae from the Client by the Contractor.

As with most cases where a simple principle is developed for payment of an organisation or individual a great deal of human ingenuity and endeavour goes into both maximising the return and protecting the original principle. To further the reader's practical knowledge I have included examples of all the basic types of maximisation I have come across in the Labour Appendix.

All larger companies will have their own scheme and a comprehensive preamble. For completeness the Labour Appendix includes a typical Incentive Scheme preamble.

6.6.3   Termination Schemes

Termination schemes are often used on major projects where there is a necessity to carry a substantial labour force into the final stages of a contract. This can be done either by superimposing a scheme related to the programme upon the existing scheme, or by totally replacing the existing one with the programme-based scheme.

The former is more likely to be accepted easily by the workforce. The Project Manager should choose a series of critical programme times and fix sums of money to each. If the payments are not progressive there will be no belief that, in say three months, the money will actually be there. This does not preclude making the last payment the largest.

The Project Manager can fund this scheme by savings against his current agreed financial budget by earlier completions. Hence by turning savings on plant, staff, service, labour and site establishment into increased bonus payments he can assume timely completions in what would otherwise be difficult circumstances.

The scheme should be introduced in sufficient time for there still to be flexibility in it, as it is impossible to allow extra time for plant breakdown, bad weather etc., which would not be self-funding. Unless there is at least three months covered there is unlikely to be sufficient programme duration for savings in time to be able to generate the funding for the additional bonus.

The application of the scheme should be as wide as possible and include all sections of the site at each date. Hence, where there are four sections there should be four items requiring completion by the first date for the stage payment. It is also wise to include in the system the statement that, if any intermediate dates are not met, the monies will be carried forward to the end payment, providing that the final date is met. Otherwise one failure at an intermediate stage will jeopardise commitment to the scheme.

6.7   TYPICAL INDUSTRIAL DISRUPTIVE TACTICS AND COUNTER MEASURES

6.7.1   Introduction

In this section the causes of the dispute are ignored and the measures and methods used by the workforce are investigated. Negotiations in a frank manner should take place at all stages to avoid industrial action. It should be realised that a large workforce is without mercy and does not appreciate generosity, which is considered a sign of weakness. The key to dealing with all industrial disputes lies in the grievance procedure and the disciplinary code which is contained in the Labour Appendix.

The crucial point is that it is recognised that, in order to maintain good morale, the employer has the right to discipline those who fail to make appropriate use of the disputes procedure in without recourse to strike or other industrial action.

It is seldom that the labour force has the patience to operate the system fully and action nearly always takes place in contravention of the agreement.

The Project Manager must realise that the most costly type of industrial action rises when the Contractor continues to pay base wages and maintain plant on hire. The pressure is all on the Contractor who is losing time and pumping out money while the operatives are not feeling the pinch. It is essential to minimise the cost, and as soon as the procedures in the Working Rule Agreement allow, cease paying the workforce and off hire the plant. This principle runs through the cases discussed below.

6.7.2   A Strike by All of the Workforce

This is the simplest form of action and the least costly. The Project Manager has only to off hire his plant and put his staff to work on planning, programming and costing exercises on a 40 hour week and continue to negotiate at any time with full time officials and stewards as necessary.

In dealing with strikes the Project Manager should bear in mind the time lag in payments. Wages are normally paid a week in hand and bonuses two weeks in hand. Hence the following examples apply:

1. If work stops at the end of a week the pay pattern is this:
    *   At end of week 1: Wages from last week's work plus bonus from previous week (i.e. a full week's pay packet);
    *   At end of week 2: Bonus from last week's work (say one third of a week's pay).

It is therefore at the end of week three that the expected pay packet is not there and the pressure on the individual is equal to that on the company.

2. If work stops halfway through a pay week:
    - After 0.5 week Wages from previous full week and Bonus from two weeks back (i.e. a full week's pay packet);
    - After 1.5 weeks Wages from 0.5 week plus Bonus for last full week (i.e. 0.75 of a normal pay packet);
    - After 2.5 weeks Bonus for last week (about 1/6 of a normal pay packet).

Hence in this case it is 3.5 weeks until the pay packet is zero. It can therefore be anticipated that, when a strike occurs, it is going to be at least two weeks before the workforce will be realistic and genuine negotiations take place.

6.7.3   A Strike by a Portion of the Labour Force

As all categories of labour are obviously necessary, and it would be very unusual for those remaining at work to carry out the work of those on strike, a costly imbalance will soon occur. The Site Project Manager should consider Rule XA(2) (d) which states: "If, in a pay-week collective industrial action of any kind, in contravention of the Constitution of the Board of this Agreement is taken by operatives employed on the site under this Agreement, the employer shall at all times use his best endeavours to provide continuity of work for those operatives who are not involved in such action and who remain available for work. In the event that, by reason of such action the employer cannot provide such continuity of work, the guaranteed minimum shall be deemed to be suspended until such time as normal working is restored."

The Project Manager should endeavour to keep the situation balanced and as soon as is possible, consistent with the obligations in XA(2) (d), he should lay-off some members of other sections of the labour force without pay. This will normally have the effect of totally closing the site, which will allow the Project Manager to off hire the plant and exist at minimum cost levels until negotiations bear fruit and full production can restart. The trades who are not in dispute will put added pressure on those in dispute to settle with management.

6.7.4   A Go-slow by the Workforce

The Site Project Manager cannot tolerate this type of action and whilst still negotiating should state that: "Unless the workforce earns its wages and complies with the grievance procedure, then the company will cease paying them. This will take the form of a verbal warning issued to the Stewards and then confirmed by a duplicated copy with each individual's name on it. Thereafter, within half a day, a written warning will be issued to each individual after which, failing compliance, the operatives will be suspended without pay." This action is entirely within the

Working Rule Agreement, and ensures that the Company's costs are minimised quickly and the pressure is put firmly on the operatives.

6.7.5   A Go-slow by a Portion of the Labour Force

The procedure applied above should first be implemented for the section of the workforce in dispute, and then the remaining section can be laid off without pay. Although this may seem hard it is essential so that trade after trade, or group after group, is not allowed to disrupt the working of the site. In pursuing these moves in the industrial action chess game, management must ensure that it is maintaining a firm but fair position in the original cause of the dispute.

6.7.6   Overtime Bans

If all work is able to be completed within a 40 hour week then this is actually cheaper and not a problem for the Site Project Manager. However, the weapon is normally only used where overtime is a regular necessary occurrence such as concreting a particular steel shutter which is moved once a day. By preventing concreting the cycle time and the unit costs are doubled. This cannot be allowed to continue and, unless a movement of hours, by putting the concrete gang on an eight hour shift starting halfway through the day works, the position must be challenged.

This can be done by the Project Manager stating that an overtime ban combined with a refusal to work staggered shifts is "collective industrial action". It is unlikely that the grievance procedure has been exhausted so the disciplinary code can then be applied and the guaranteed minimum withdrawn.

6.7.7   Summary

To summarise this seemingly bellicose section it must be made clear that if the site is well planned, discipline is fair and clearly established, disputes will be greatly reduced. It is inevitable that on large sites, with directly employed labour, the Site Project Manager will be tested and he should respond firmly, making sure that the Company's position in the dispute is entirely reasonable before taking the measures suggested.

CHAPTER 7

Insurance

7.1   INTRODUCTION

In this Chapter I deal with the mechanisms of the policies and the recovery available to the Contractor. The contractual requirement to insure and the amounts are within the Contract Appendix and it varies with the various Conditions of Contract.

7.2   CONTRACTOR'S ALL RISK POLICIES

Considered now are the basic provisions of the Contractor's "All Risk" Policies and the way they operate.

On major projects a specific policy will be taken out for that project recognising the particular risks involved. Minor projects will normally be covered by a "floater" policy, either by type of work or specific time periods covering all the Company's work or both.

There will be a number of differing excesses dependent on the type of risk and these should be studied carefully, before making a claim, as a loss adjuster will dispute the nature of the incident to try to apply several or higher excesses. For example, in one case divers had failed to clean out the returns in a sheet pile wall and a tremie concrete plus was weakened, a series of blow-outs occurred in the plug. Although there was only one incident of bad workmanship, the insurance company endeavoured to claim that as there were several blow-outs therefore several excesses applied.

All policies exclude consequential loss, hence the Contractor will not recover the cost of lost time and liquidated damages; the extended overheads often being the most costly effect of the damage. As is covered in the analysis of the Conditions of Contract, liquidated damages should be avoidable unless the damage is due to fault or negligence on the part of the Contractor. The Contractor should seriously consider enhanced spending to avoid time loss (apart from the fact that it is usually cheaper anyway) because direct expenditure will normally be recoverable under the policy whereas expenditure from being on site longer will not.

The direct results of faulty design and bad workmanship will be excluded, but the consequence of that bad design or workmanship on other parts of the structure or temporary works are covered. This is entirely consistent with the requirement in the contract for the works to be insured against all eventualities. For example, if a base is faulty and has to be replaced, the cost of removing and replacing the column constructed on top is covered, subject to the policy excess, but obviously not the base itself.

Some policies contain a clause which specifically states that expenditure to prevent or minimise loss or damage will be covered up to a value that would be the notional cost if the damage had not been prevented. This is an unusual clause and most policies are only activated by actual damage. The following points should be borne in mind:

1. It is a principle in insurance that the policy is only activated in workmanship cases where the Contractor had "achieved his objective". For example, an insurance claim would be rejected for damage caused by water leaking out of a pipeline which had never been completed and had been backfilled in error. The same damage caused by a leak in an incorrectly assembled but completed main would be covered. The reason for the difference is that in the first case the work was incomplete whereas it was complete in the second case.

2. The Contractor should study the particular policy and remember that what is reasonable will normally prevail. It would be extraordinary if an insurance company admitted liability immediately.

3. The insurance company should be informed as close to the incident as possible and even before the Contractor is certain that it is covered by the policy. The Company can always withdraw if it becomes obvious, as the incident develops, that it is outside the policy.

4. With "floater" policies (for instance, on Joint Ventures where a separate company is formed) relating to a specific time period, notification must be made before the time period expires or the new policy will not cover the situation. If the Company should have known of the incident at the time the previous policy expired then they may be time barred.

7.3   PROFESSIONAL INDEMNITY

As the contract provides for payment of the Contractor's costs, consequential and direct, for the errors of the Engineer, it is of passing interest to note that the Engineer will be covered by Professional Indemnity insurance against claims arising from his acts or omissions. Subject to a fairly large excess it covers the costs incurred by the Employer under the Contract as a result of any errors by the Engineer, and other external individuals and bodies not party to the contract. Also

Professional Indemnity Insurance cover depends on the Contractor/Consultant/ Subcontractor being insured at the time the claim arises, not when the faulty design was incorporated into the works.

Of greater relevance is the fact that "in house" design departments as well as subcontracted external designers and consultants, must be covered for the consequence of faulty design or advice.

Hence, a faulty temporary works scheme at tender stage executed by one's internal design department could well lead to a claim for reimbursement of any extra cost incurred in producing a working design from the Insurers. The inadequacy would, in my opinion, have to be of a structural nature rather than one that only adversely affected production.

Therefore, if a design fault in temporary works causes loss, the Contractor can look to the design department's professional indemnity insurance for the cost of rectifying the defect and the Contractor's all risk policy for the effects, if any, of the defect on other parts of the structure.

7.4   ACTION GUIDANCE

The Contractor is required to build the works quickly, efficiently and in a safe and secure environment. Insurance cover is there as a back-up to failure. As every motorist knows today's claim is tomorrow's premium increase.

Most cases of vandalism or minor damage during the course of the works are close to the excess and although disruptive (a consequential cost not covered) will not be worth claiming.

The Contractor must be aware of the parameters so that if he chooses not to initiate a claim it is after having evaluated the likely recovery.

The following guidance procedure for damage to the works should be followed through. It is a summary of advice in this section:

1.   Ascertain the cause of the damage.
2.   If it is probably an excepted risk under the contract, the Contractor should notify both Engineer and Insurers. There is no need to tell either that the other is being notified. However, it is fraudulent eventually to obtain monies for the same costs from two sources. Excepted risks bring payment for consequential costs, whereas insurance does not.
3.   If the damage is to subcontract works, the Subcontractor on the FCEC form of subcontract is liable and should be instructed to put the works right. The Contractor's own insurers should also be informed.
4.   If faulty workmanship or defective temporary works design has led to additional costs, the Contractor should try to define a point at which the incident finishes and the damage commences and inform the insurers.

5. In all defective temporary works designs look to a holder of professional Indemnity Insurance, which may be an outside consultant or the Contractor's own policy.
6. In all cases of loss or damage to plant or equipment inform the Insurers. The excepted risks still apply to this damage.
7. Whenever possible obtain or apply for a completion certificate. This is particularly important on multi-contractor sites and where vandalism is a problem.

CHAPTER 8

Contract

8.1 INTRODUCTION

As a result of giving courses over a period of 20 years on this subject it has become apparent that less information is more effective than saturation. For maximum impact I have concentrated in this Chapter on the "Formation of a Contract" (which applies equally to the Main and Subcontracts) and the "Key Commercial Actions during the Contract". The latter are universal and applicable to all the various Main Contract Conditions.

Detailed consideration and comparison of the various Conditions of Contract are then contained within the Contract Appendix, which is designed both as a guide to the various conditions and a means of comparison to enable someone familiar with one set of conditions to use another.

8.2 FORMATION OF A CONTRACT

Major public works contracts will always have a full formal contract in place before any work is executed. In some cases on small projects a formal contract might not be in place where the requirement for the work to be executed has often been discussed over a period and suddenly becomes urgent in the eyes of the client. The criteria are the same for formation of a Subcontract where the Project Manager must view himself as the Client.

For there to be a Contract there must be:

- an offer
- an acceptance
- intention to create legal relations
- consideration.

None of the above need be in writing, but an oral contract is very risky and does not come within the Housing Grants Act which permits Statutory Adjudication within the UK. Written confirmation by one party of oral acceptance will suffice to have a contract subject to the provisions of the Housing Grants Act.

In almost every case involving construction there will be an intention to create legal relations. This caveat covers offers such as an agreement to meet at a set place and time. Consideration was defined in 1875 in Currie v Misa as follows:

> "The essential principle of consideration is that **there must be a reason why each promise is made**, in return for which the promise is given. The promisee must **give something** or **do or forbear from doing something** in return for the promise made to him, in order for his own promise to be legally enforceable. Consideration involves either **some detriment** to the promisee **or some benefit** to the promisor: 'A valuable consideration, in the sense of the law, may consist either in some right, interest, profit or benefit accruing to one party, **or some forbearance, detriment, loss or responsibility given, suffered or undertaken by the other'**."

Consideration on this basis always occurs when one party mobilises to site and this is why the last document in place when this happens tends to determine the conditions under which the work will be deemed to be carried out when no agreement is reached.

The problems associated with formation of a contract can be illustrated by Site Investigation contracts which are carried out on ICE-based contracts.

A Consultant can be developing a scheme with an Entrepreneur when they have no formal contract between them. It may not be decided which of the Entrepreneur's shell companies is going to be the Employer.

The first ground rules before commencing are:

- who is the Employer?
- do we want to be in Contract with him?
- does the Consultant have power to commit the Employer?

The next set of ground rules relate to the Conditions of Contract that will be deemed to apply. This is necessary because the last correspondence on file before commencement will almost certainly determine what will hold sway. An order or confirmation with different terms can be held to be a counter offer which without demure will be deemed to be accepted by performance.

- Where the Employer (or Main Contractor) has not issued formal order or contract, **confirm that you are starting on your conditions and quotation**. A contract is then formed.
- If an order is received with different Conditions these must be checked for acceptability.
- If an order or confirmation of acceptance is received which refers to any other conditions these must be obtained and checked for acceptability.

The parties are at liberty to formalise their contract after commencement and discussions may continue; it is wise for site personnel to avoid this. It is sensible for the company to do this under the caveat of "without prejudice" because another criteria for determining what the contract contains is to examine when the parties were closest together (consensus ad idem). This process inevitably means you will have made concessions and you may not have received sufficient benefit. (Also you will not have your Director's approval).

8.3   KEY COMMERCIAL ACTIONS DURING THE CONTRACT

8.3.1   Confirmation of Instructions

Without confirmation the issuer of the instruction may well say later that what was said was advice or an additional option. **Without confirmation the Contractor may well not be paid.**

8.3.2   Notices

From a non contractual point of view it is obvious that any purchaser of any service expects the price to be what he originally calculated or was informed unless he is told otherwise.

In the review of the different Conditions of Contract in the Contract Appendix it will be seen that the lack of notice can eliminate and/or seriously reduce the Contractor's recovery.

To be able to give notices when they are required all the site personnel need to know the expected productivity for each operation under their control and the conditions and access entitlements under the contract.

The main reasons that such prominence is given to notices are twofold:

1.   Notice is a warning and it permits the Employer to consider or take alternative action to avoid or mitigate the perceived delay and or additional cost.
2.   It permits the effects to be monitored, verified and agreed.

Notices must be in writing and preferably not email unless this is specifically the agreed form of communication. Fax has been established as good service but there is no such precedent for Email although the 7th Edition of the ICE Contracts now provides for this means of communication.

Notices should be factual and avoid emotive language. They should clearly identify the event or future event and if appropriate seek instruction. It is not necessary to identify the contractual clauses, either under which they are given or under which additional time or cost is sought.

By far the greatest cost to any Contractor is the cost of being on site. In every notice given it should be considered whether there is reason to ask for an extension of time.

For a summary of the matters which are the Employer's risk and for which the Contractor (and his Subcontractor) should give notice see Chapter 9, Section 9.2.

Subcontractors also need to give notice when the Contractor does not provide access in accordance with the Subcontract or fails to provide the agreed attendances or services.

8.3.3   Records

Although tedious, lack of records equals lack of recovery of the Contractor's full entitlement and potential losses, which negates the whole purpose of the contract.

Hamish Lal, writing in "Quantifying and Managing Disruption Claims", makes the following point.

> "By now, nearly all practitioners are aware of the importance of site records in retrospective delay and disruption analysis. For at least the last 20 years, the relevance of records has been stressed, for example, Abrahamson:
>
> **'A party to a dispute, particularly if there is arbitration, will learn three lessons (often too late): the importance of records, the importance of records and the importance of records.'"**

Records lose some of their potency when notice has not been given that they are being kept and so it is essential to do this at the time.

Records need to identify the individuals and resources, their location or activity and its duration. It is best that they should be signed daily but if the certifying party will not co-operate a notice that they are being kept will sffice.

8.3.4   Measurement of the Work

Where the Employer has had the Works designed the Specification states how any particular item is to be executed. The Method of Measurement provides what is deemed to be included in any particular price. As part of measuring the work the site should check that the bill descriptions are appropriate. The Engineer can and does make errors in his selection of items to price. New or additional items and more favourable rates should be sought wherever appropriate.

CHAPTER 9

Claims

9.1   INTRODUCTION

As with all sections of this book, this chapter seeks to introduce the reader to the main principles and actions necessary, and give general guidance as to how to prepare submissions.

The headings under which the analysis is conducted are firstly "Claim Categories" which define the different bases on which claims are founded. This is followed by "Contractual Principles Applicable to Most Claims".

This chapter concludes with the "Format, Presentation and Content of Claims" and "Negotiation of Claims".

Before commencing the first section it is as well to first define a claim. Claims may be of various kinds, for example for extra payment, for damages, or for extensions of time, and may arise either pursuant to express powers in the contract or for breach of the contract. Typically, claims are the subject of a formal claim document produced by the party making the claim, upon which a decision is given under the relevant contract or negotiations take place between the parties and their advisers, followed by adjudication, arbitration or litigation in the event that settlement proves impossible.

The second action is to consider how to avoid claims. A claim is something which has become a potential dispute and which requires a formal position statement. Any extra under the Contract under any clause from Unforeseen Conditions to Variations, together with the full associated effects, can be negotiated at the time and need not be a claim.

Avoidance of claims does not mean avoiding increasing the payment under the Contract; it means avoiding conflict and disputes.

The prerequisites are:

1. Good personal relationships with opposite numbers which must be worked at.
2. Knowledge of the technical imperatives as well as the contractual entitlements.
3. Early notices with constructive suggestions to minimise effects.
4. Accurate assessment of effects. Do not be over optimistic in the early days and absorb or minimise early delays.

5. Accurate forecasting of the potential cost so budgets are prepared and options examined and the sum quoted is not exceeded by the event as it unfolds.

The principles laid out in the following sections apply to claims made by a Subcontractor under the Subcontract or a Main Contractor under the Main Contract.

9.2   CLAIM CATEGORIES

9.2.1   Measurement Claims

These obviously only apply to contracts where there is a Bill of Quantities and Method of Measurement. There must always be a reason for rerating of the work, other than that of a loss being made. Contractually, the Contractor can be forced to do additional work at a loss. The relevant two sentences in the ICE-based Conditions come from Clause 52 (1):

> "Where work is of similar character and executed under similar conditions to work priced in the Bill of Quantities, it shall be valued at such rates and prices contained therein as may be applicable. Where work is not of similar character or is not executed under similar conditions the rates and prices in the Bill of Quantities shall be used as the basis for valuation so far as may be reasonable failing which a fair valuation shall be made."

If the Contractor is making a loss on "relevant rates" his only course is to concentrate on investigating and demonstrating any differences of timing, rate of working, numbers or uses of temporary works items and the like. This latter category will be most often connected with reduced quantities and Clause 56 (2). (Four examples are given in the Claims Appendix.)

The tender will govern whether the Contractor's efforts are directed towards a rates-based exercise or convincing the Engineer that a separate "fair valuation" based on the recorded cost of executing the works should be made.

Measurement claims are also applicable where non-standard items or details are included in the Contract without proper description or alteration to the Method of Measurement in the Preamble to the Bill of Quantities.

9.2.2   Claims Under the Contract

This is where the majority of claims are founded and the principles are recognised in the various Conditions of Contract. These principles are not included in every Contract. The Reader will have to consult the Contract Appendix to see if the particular Conditions of Contract includes the principle as a potential risk borne by the Employer.

9.2.2.1 Primary Causes

- Compatibility of Contract Documents
- Further drawings and instructions necessary for construction of the works
- Unforeseen conditions
- Method related changes
- Impossibility
- Provision of permanent works' design criteria for temporary works design
- Suspension of works
- Variations
- Possession of site
- Delay in Communications (NEC only).

9.2.2.2 Secondary Causes

- New statutes
- The works or instructions being in breach of statutory provisions
- Named third parties' requirements not conforming to the relevant contract special requirements, statutes or other regulations
- Street works Act: new conditions or restrictions post award
- Nominated subcontractors:
  a. conditions for entering into subcontract:
  b. default.
- Damage exemptions:
  a. use or occupation by employer;
  b. fault or error in design by Employer's agents;
  c. inevitable consequence of executing the works.
- Tests proving that materials or work complies with requirements.

9.2.3 Global Claims

Often practitioners consider these to be a separate class of claim. They are only a particular method of valuing the costs of a series of effects with different contractual causes and entitlements in a single rolled up evaluation which is often based on the total increased cost.

They are to be discouraged. The Case Law principles surrounding Global Claims are one of the topics in the Claims Appendix but the key requirement for any chance of success is that no element of the increased cost has been caused by matters which are the Contractor's risk.

9.2.4   Set off and Claims against Subcontractors

In the UK the Housing Grant and Regeneration Act (HGRA) has opened up the contractual relationship between contractor and subcontractor to far greater scrutiny, requiring the Contractor to be far more careful in his claims against the Subcontractor during the course of the contract.

Subcontractors are now scrutinising reductions in their applications to see if the Act has been breached and cash flow can be improved by use or threatened use of the Act. To be entitled to set off any monies at all the Contractor must have gone through the sequence of producing a statement of the monies deducted after the Subcontractor's application and before payment is due, as particularised in Chapter 5 and the Subcontract Appendix.

Reductions of a Subcontractor's account can be divided into two parts which are:

1.   Reduction of quantities or items claimed;
2.   Counter claims.

In the first case, the measurement under the Main Contract by the Contract Administrator should be sufficient proof that the reduction was warranted. A copy of the marked-up valuation will suffice as documentary evidence. Similarly, where back to back extras are claimed it will be sufficient to red line these with comments such as:

- insufficient particulars;
- under consideration by CA.

Extras against the Contractor for which he has no recourse under the main contract can only be rebutted by a. above. Once detailed particulars have been provided they must be considered and reasons for rejection in whole or in part given in the set-off notice.

Direct additional costs sought by the Contractor from the Subcontractor must have sufficient details to identify the cause, entitlement under the Subcontract and individual elements of the charge. Unparticularised amounts will not be upheld at Adjudication.

The more difficult case is for counter claims for delay and/or disruption. Particulars in the interim which should be provided to avoid challenge at Adjudication on the basis of procedure alone are:

1.   Proof of the original programme requirement;
2.   Proof of delay;
3.   Notice of delay;
4.   Notice of potential effects and resources affected;
5.   Particulars of the costs associated, together with warnings regarding potential further costs not yet incurred (for example, L & ADs, other

subcontractor's prelim claims, (however these cannot be set off until they have been incurred).

The final claim against a subcontractor should contain the same level of detail as if it were made against the Employer.

9.2.5   Claims Outside the Contract

The other types of claim that are rarely incurred and are detailed in the Claims Appendix are Restitutionary Claim, Breach of Implied Terms, and claims under the Law of Tort.

9.3   CONTRACTUAL PRINCIPLES APPLICABLE TO MOST CLAIMS

9.3.1   Introduction

One of the more annoying aspects of modern life is the presentation of information under the heading of "Your Questions Answered", which are not what you would have asked. I am also going to cover a series of principles which constantly reoccur. One of the best sources to quote as an authority for these principles is the Society of Construction Law's Protocol which is available on the Internet.

9.3.2   Records and Resources

The Protocol's draft clauses for records show what could soon become the norm. To ensure that all costs are picked on in retrospective valuation of variations and claims a resourced As-built programme is required. Practically the resources will probably be in spreadsheets whilst the programme will be in Primavera or similar software.

      The cost centres used should recognise the way the project is being managed which should match the original programme. The use of the figures is discussed later.

9.3.3   Lack of Proper Notice

The only contracts which actually state that Notice is a condition precedent to payment are FIDIC and the CECA Blue Form subcontract conditions. All other contracts provide notice procedures which must be complied with and penalties for not doing so which potentially result in a lesser assessment than would otherwise have been the case.

The most stringent are in the Highways and New Engineering Contract. Here the intention is to have the full effects established using the accepted programme at the time. Whether a claim can be ruled out entirely upon lack of notice under FIDIC will ultimately depend upon the applicable law referred to in the Contract.

Within the UK the more likely cause is between Contractors and Subcontractors. The current situation is admirably explained in Hamish Lal's "Management of Disruption Claims". In summary, the current position is that the "Prevention Principle" means that a defendant cannot benefit from non fulfilment of a contractual duty which would be the case if he succeeded in having a claim struck out on the grounds solely of lack of notice. That said, there is a Scottish case to the contrary and Subcontracts under Scottish law may be more vulnerable to this approach.

9.3.4   Float in Programmes

An extension of time is only necessary when Employer's risk events reduce to below zero the float on activity paths affected by the delay. In principle, where the Contractor intends to complete the works before the Contract completion date and is prevented from doing so by the Employer's risk events, then he is due compensation for the time-related cost directly incurred despite there being no delay to the Contract completion date.

9.3.5   Concurrency of Events

Where a Contractor's risk event occurs concurrently with an Employer risk event the former should not affect the time due for the latter. However, only the costs resulting from the latter are due and a split of resource on the respective items will be necessary.

9.3.6   Mitigation

Unless there are express provisions to the contrary, the Contractor has only a general duty to mitigate effects on the works of Employer risk events. It does not extend to having to produce additional resources (see Acceleration below).

9.3.7   Acceleration

Where there are no provisions in the Conditions of Contract for ordering acceleration (principally the ICE 5th) the measure should be agreed and the basis for payment agreed.

Although undesirable, there will continue to be claims for "Constructive Acceleration" (where lack of a granting of an extension at the proper time leads to measures to complete to the original date) if Contract Administrators do not promptly and properly assess events notified to them and extend time for completion at the time.

9.3.8 Evaluation of Extended Time

The principle is that the Contractor's actual additional costs should be reimbursed for any particular agreed event. These are the costs at the time of the event and not at the end of the project when the overrun occurred.

9.3.9 Head Office Overheads

The use of formulae for recovery of Head Office overheads as a result of contracts being prolonged by Employer's risk events has been supported with most tending towards the Emden Formula, which relies upon actual overhead in the period.

Although the proponents of the formula do not mention reductions to the sums generated if there are variations taking the total paid above the Contract Sum, the Society for Construction Law recommends that it should be discounted when these are above 10% of the Contract Sum.

The formula is:

$$\frac{\text{Contract Sum x Overhead and Profit \% for relevant year x Extended Period}}{\text{Contract Period}}$$

9.3.10 Claims for Interest or Financing

These are generally supported and can only run from when the Contractor actually asked for the sums in question. The backstop position is that it runs from when full particulars have been provided. Each case will be different depending on what was obvious and how pedantic the requirement for detail has been. Interest is due separately in the Adjudication process.

9.3.11 Costs for Preparation of Claims

Generally these are only recoverable when the Contract Administrator had sufficient details at the time and the entitlement was obvious. However, costs for external experts required in the preparation of submissions which are successful in whole or in part are normally recoverable in any case.

9.3.12   Methods of Evaluating Delay and Disruption

There are three aspects to this which are:

- justifying the actual prolongation period;
- valuing the cost of prolongation of time-related resources; and
- assessing any reduced productivity.

The first two have to be programme based. Indeed, there is Case Law which states that if the Contract Administrator does not use a programme based method his assessment will be held to be fundamentally flawed. There are the following programme options identified in the Society of Construction Law's protocol:

- as planned compared with As-built;
- impacted as planned programme where the additional durations for the events are inserted into the original network;
- collapsed As-built programme where a detailed As-built programme with the actual logic and duration is produced and the actual duration of Employer's risk events deducted;
- Time Impact Analysis: this involves inputting each delay into the actual programmed intentions at the time. Vastly costly and complicated; only in circumstances where the New Engineering Contract provisions for constantly upgraded programmes for each and every Compensation Event would it be likely for sufficient record or detail to exist to achieve this.

The difficulty of using the Collapsed As-built method is that it requires agreement between the parties to create the actual As-built network. It is difficult to produce precise points for the demand lines between activities and a further source of dispute is engendered.

Operating in this field for 20 years, I have found that the most realistic results for a reasonable cost come from the use of a modification of the Impacted as planned programme compared with the As-built programme. In most cases where there has been considerable disruption by multiple Employer's risk events the Contractor has sought to mitigate by departing from the original intention of sequential working and starting subsequent operations in parallel with overrunning operations. Thus, when the events are input into the original logic it is often the case that far more time is generated than was actually taken. By inspecting the As-built programmes I determine the actual mitigation measures of overlapping activities undertaken by the Contractor and apply them to the Impacted Programme.

In most cases this then demonstrates a good fit between the Mitigated Impact programme and the As-built programme providing acceptable evidence that the periods taken to undertake the various sections of the works were dictated by the Employer's risk events.

Valuation of core preliminary resources can be made on the overall additional period and sectional preliminaries attached to specific work areas similarly evaluated.

There are two basic methods of evaluation disruption, the first of which is the measured mile approach which compares labour and plant earnings ratios at Contract rates both before and after disruption. The disadvantages are that there may not be sufficient or relevant samples of work pre-disruption to cover all categories of work.

The second method is based on the prolonged envelope of resources connected to any particular series of items that has been justified by the programme exercise described above. This presumes that the resource for any item remains roughly constant. Allowances would need to be made if the Employer's risk events included significant additional quantities which had resulted in adequate remuneration of direct cost.

9.4   FORMAT, PRESENTATION AND CONTENT OF CLAIMS

9.4.1   Descriptive

There are some basic principles to adhere to. The first imperative is to tell the reader what the dispute is about. The senior decision maker may well read only this section so brevity and clarity are called for in either a Précis or an Introduction. This should summarise the contractual basis of the claim, the sums and the time sought. It is worth assuming that every document has the potential for Adjudication and there should, therefore, be a section on Contract Particulars giving details of the parties, the Contract form, period for completion, etc.

For clarity the Narrative should be uncluttered by continual reference to the original programme intentions and any particular contractual arguments. It is, therefore, better to put these in earlier sections so that the reader is given a clear view of these before he has to grapple with the events.

The narrative's principal function is to particularise how the events unfolded, to demonstrate that adequate notice was given and to identify what is to be valued and why. It should refer to the documents that prove its contentions, which can either be presented as part of the text or held in appendices.

For a Delay and Disruption claim the narrative will speak to a programming exercise of the type described in 9.3.12. The events will be held, together with the relevant back up, in Impact or Change Notes, which describe the effect on the particular elements of the programme, in appendices.

The effects are then listed in the order of preparation in spread sheets, sorted by programme activity and summed to give a total to be input into the programme. The narrative will typically identify the key items on the critical or sectional critical path and detail the prolongation of core and sectional preliminaries and the periods at which it took place for valuation in the Quantification.

9.4.2   Quantification

The order I normally employ is to commence with any direct costs of particular events that are disputed. This is followed by preliminaries split into:

- core preliminaries;
- section preliminaries; and
- additional preliminaries.

This would be followed by specific direct cost items and the disruption or loss of productivity calculations based either on the measured mile or the prolongation of envelopes of resource derived from the programmes above.

I follow this with additional off site overheads for the overall period and financing costs. All details can be held in appendices. Where the project is large and the sums claimed are also large, the appendices can contain extracts from the cost systems provided that this identifies the invoices or other source documents which can be subsequently checked.

9.5   NEGOTIATION OF CLAIMS

A good starting point, whether trying to persuade a client or compromise with a subcontractor, is to invite the other side to consider the process that would be entailed if the matter was taken to Court or Arbitration.

1.   Accuracy of the particulars of the events will be supported by witness statements which will be the basis of the case in the tribunal.
2.   A programme expert would be appointed to consider whether the alleged events had or should have had the effects alleged.
3.   A cost expert would be appointed to check the costs arising from each alleged effect to the extent that they are actual, reasonable and attached to the event. He will rely on the Programme Expert's view of the proper periods.
4.   The notices will be examined to ensure that the defending party had the opportunity to take different action where this might be possible. If Notice is a condition precedent this will be central.
5.   The contractual or other entitlement will be analysed.

Without prejudice parallel discussions on these five principles should be encouraged, irrespective of whether there is agreement on item 5. Liabilities are established aspirations suppressed and both parties understand the position better.

CHAPTER 10

Adjudication

10.1   INTRODUCTION

Within the UK the introduction of Statutory Adjudication, whether or not there were provisions in the contract for it, has revolutionised the way disputes are dealt with. Because it is quick and potentially as cheap as Conciliation or Mediation it has pushed these Alternative Dispute Resolution (ADR) methods to one side.

It has the considerable advantage that a third person takes a view which means that the senior people on each side do not have to report that they have agreed to a different outcome than they predicted and accept that they were wrong. Mediation is now only regularly used as a precursor to court action, which is obviously after Adjudication.

Because it has become so widespread I have included this chapter on the subject but have not touched on Dispute Resolution by Arbitration or the Courts, which is almost entirely controlled by Lawyers and will not normally be experienced by a construction manager.

Essentially, Adjudication was designed as a quick fix (although it must not be entered into lightly) to release cash flow; the parties are then free to take the matter to Arbitration or the courts thereafter.

Practically, the majority of disputes are now settled at Adjudication in the UK including complex issues. In the latter instance an increasing use of lawyers has meant that it has almost become fast track Arbitration for the more complex issues.

The purpose of this chapter is to introduce the reader in to what to expect during an Adjudication; it is not intended to be a comprehensive work that will permit all permutations of events to be covered. However, a case prepared in the manner described in this chapter should be in sufficient detail to accord with the Act.

Because almost all parties within the construction process have the right to adjudicate (see below) everyone within the construction process might at one time or another be a Defendant or Claimant.

After "The Right to Adjudication and Limitations" the necessary actions by both parties in an Adjudication are examined to accommodate this fact and particularly to provide details for a defendant.

This Chapter is completed by an analysis of the Adjudication process and the bases upon which an Adjudicator's decision can be challenged.

10.2   THE RIGHT TO ADJUDICATE AND LIMITATIONS

This is considered in three parts:

1. What is permitted under the Act;
2. When a dispute is deemed to have arisen;
3. Limitation to the process.

A Construction Contract is defined in Sections 104 and 105 of the Act as an agreement to undertake the following operations:

- construction, alteration, repair, maintenance, extension and demolition or dismantling of structures forming part of the land and works forming part of the land, whether they are permanent or not;
- the installation of mechanical, electrical and heating works and maintenance of such works;
- cleaning carried out in the course of construction, alteration, repair;
- extension, painting and decorating and preparatory works;
- contracts with architects, designers, engineers and surveyors as is the giving of advice on building, engineering, decoration and landscaping.

There are exclusions from the Act which are likely to be encountered and these are:

- supply only contracts where no installation takes place (i.e. aggregates or reinforcement for site fixing);
- extracting natural gas, oil and minerals;
- work on process plants and supporting access steelwork where primary activity is nuclear processing, power generation, water or effluent treatment, handling of chemicals, pharmaceuticals, oil, gas, steel and food and drink (excluding warehousing of these). Cases coming to the Courts to clarify the exclusions under this heading suggest that the basis is whether the particular contract works are part of the process;
- contracts with residential occupiers;
- PFI contracts applying to the heads of agreement, contracts for finance, insurance policies and bonds and agreements under some statutory provisions;
- contracts not in writing: provided that there are letters, even from a third party, defining what is to be done it will most likely be deemed that there is a contract in writing;
- matters before the contract came in to existence, such as any misrepresentation which might have induced one party to enter into it. (The Adjudicator is permitted to decide upon the correct interpretation of the terms of the contract);

- contractual or legal advice regarding construction contracts;
- where a Contract Administrator has a set period for an action or decision, a dispute cannot arise on that aspect until it has elapsed.

The second element is whether a dispute has actually arisen. Many standard forms of contract, some of which have been reviewed in the Contract Appendix, have endeavoured to provide a procedure to be followed before a dispute can be deemed to have arisen. These will not be upheld where they appear to prevent the claimant from taking a matter to Adjudication at any time.

However, Case Law has established that the Defendant must have had adequate time to consider the documentation (the more complex the longer required) and have declared his opinion and that the Principals should have met to try to resolve it. (This means Director level contact.) The Claimant should always give a period for a response to any submission where there is not a proper procedure in the contract.

The third element is the limitation of argument or content. Anyone familiar with law will know that cases can turn on new facts and/or expert opinion. No new facts, opinion or other data can be introduced into the Adjudication process once it has started.

This is most important for a Claimant who is eager to commence the process and receives a reply to his submission with new data in it. He should reply to these facts (if they are relevant) before he initiates the Adjudication procedure.

Where there are technical or legal issues involved expert assistance should be called upon before the Adjudication procedure is commenced to avoid falling foul of this third element.

The second part of this element is that only one dispute can be referred at any one time and if the sums involved relate to more than one issue it should be termed in the Notice of Adjudication as a dispute regarding the payment of monies.

10.3   ACTIONS FOR A CLAIMANT

Where the contract includes its own Adjudication rules, and these are deemed to be in accordance with the Act, with possibly a named Adjudicator, the case proceeds in accordance with the provisions.

The following considers a case under the Act. The Claimant issues a Notice of Adjudication which must contain the following and be consistent with the matters in the dispute:

- nature and brief description of the dispute and the parties involved;
- when and where the dispute arose;
- nature of the redress being sought;
- names and addresses of the parties to the contract.

The principal purpose of the Notice of Adjudication is to inform the Adjudication Nominating Body of the nature of the experience and expertise that the Adjudicator should have. If the contract does not even specify a Nominating Body then the Claimant is free to choose whichever body he thinks has the most relevance to his dispute. Once the Adjudicator has been appointed, the Claimant submits the Referral Notice to both the Adjudicator and the Defendant. The Notice should:

- be consistent with the Notice of Adjudication;
- explain the nature of the dispute and how it arose;
- detail the facts that are relied upon;
- provide the documentary evidence to support those facts;
- provide sufficient details of the contract to show that there is a contractual right to the remedy sought;
- not include evidence that the other side has not seen before; this could be challenged and possibly stop the Adjudicator's decision being enforced;
- list the decisions that the Adjudicator is required to make.

Practically, I would normally use my original submission, prepared as described in this chapter, update or insert an interest calculation, insert the Defendant's replies to any element and then add rebuttal which presents all the necessary documentation in one volume.

An additional preliminary section will always be necessary to particularise the specific decisions sought from the Adjudicator which will normally be that various events lead to entitlements to specific sums. There should be a general caveat "or other such sums as the Adjudicator may decide" to avoid a challenge that the dispute was as to whether a specific sum was due. It would be in this section that, if the contract permits, it would be possible to seek that the Defendant pays the Adjudication costs and expenses as well as an order that all awarded sums be paid in a set time.

The award must be made within 28 days of the referral unless the Adjudicator seeks a 14 day extension with the parties' agreement.

During the 28 day period the Defendant's response to the Referral will be received. This needs to be studied for any new facts, opinions or other data relieved upon. There is no automatic right of reply and new elements should be identified to the Adjudicator.

10.4   ACTIONS FOR A DEFENDANT

Having already identified that there is no automatic right of reply, some might consider it more advantageous to let a dispute develop by allowing the period for response to a claim submission to run out and then reply during the Adjudication process. This is dangerous as the response will all be new and, providing it is not ruled out by the Adjudicator, can result in him allowing the Claimant time to reply and awarding an additional proportion of cost against the Defendant. This is also a form of ambush and will be resisted.

Another good reason to avoid this is that it is more likely that elements of the Defendant's alleged costs can be part of the dispute. If computed and submitted after the Notice for Adjudication they are likely to be viewed as a counter claim and a separate dispute which will be heard after the Defendant will have had to pay any award.

The first action is to check that all the matters in "The Right to Adjudication and Limitations" have been complied with, particularly if the Defendant is surprised to receive the Notice. The Law and Adjudicators frown on ambush tactics and the Defendant will be allowed adequate time to reply, but this will be viewed in context of the Scheme for Contracts and will not be generous.

Case Law has established that there are four options open to those who challenge jurisdiction which are as follows:

1. Agree to widen the jurisdiction of the Adjudicator and so refer any dispute about jurisdiction to the same Adjudicator;
2. Refer the dispute about jurisdiction to a different Adjudicator;
3. Seek a declaration from the court as to whether the Adjudicator had jurisdiction;
4. Reserve their position, participate in adjudication and then challenge any attempt to enforce decision on jurisdictional grounds.

Challenges to the Adjudicator's jurisdiction have more chance of succeeding where they rely on a dispute not having developed and the Adjudicator will be careful in this respect. If, for whatever reason, the Defendant still challenges jurisdiction the wisest course is 4 above because the Adjudicator is likely to proceed anyway and a judgement against the Defendant could be enforced, which could no doubt have been less if the Defendant participated.

On receipt of the Referral Notice the Defendant should begin his response immediately as the Adjudicator will require it quickly (probable maximum 14 days from Referral). It is up to the Claimant to make his case and the Defendant should point out where he has not proved any particular contention.

Similarly, the Defendant must deny each particular decision sought by the Claimant or the Adjudicator may consider that element uncontested and award accordingly.

Provided the Claimant has presented his case as I suggested earlier, a Defendant would be best advised to concentrate on a succinct précis of the arguments against the matters claimed to reinforce these principles in the Adjudicator's mind.

10.5   THE ADJUDICATION PROCESS

The Act requires that the Adjudicator take the initiative in ascertaining the facts and he may or may not require a formal hearing as well as being able to:

- request any party to the contract to supply any documents the Adjudicator reasonably requires, including further written statements;
- decide what language should be used, including whether translations are needed;
- decide whether to meet parties and their representatives;
- decide whether to make site visits and inspections (the Adjudicator may need the consent of people outside the adjudication before he can do this);
- impose deadlines or limits to the length of documents or oral representations;
- issue other directions for the conduct of the adjudication.

Where the Adjudicator asks for particular documentation it is clear that the party who is asked for it will be disadvantaged if it does not supply it because the process continues as the decision must be made in the timescale provided and will be made on what the Adjudicator has before him.

It has already been set out that new opinion or data cannot be introduced into the dispute; Case Law has suggested that where the Adjudicator seeks expert opinion, and he is required to notify the parties, he must reveal the advice to the parties to enable them to answer it (RSL (South West) Limited – v – Stansell Limited (16 June, 2003)).

10.6   CHALLENGES TO AN ADJUDICATOR'S DECISION

This is not a subject for this book; however, the Courts will make every effort to uphold a decision or particular element of a decision, even if there are errors in it. Successful challenges in the Courts have been mounted where:

1. The Adjudicator has strayed outside the question posed in the Notice for Referral and did not, therefore, have jurisdiction.
2. Errors which are fundamental, such as incorrectly identifying parties or assuming that a different set of conditions applied.
3. Breaches of natural justice where documents and/or opinions obtained by the Adjudicator were not made available to the parties.

The sums in the Adjudicator's decision have to be paid and other sums cannot be set off against them.

The decision can always be overturned by taking the full case to Arbitration or the Courts, as provided for in the contract, where new facts or arguments and even entirely different grounds can be used.

CHAPTER 11

Design Management and the CDM Regulations

11.1   INTRODUCTION

The intended audience for this book is wider than the UK and, at first sight, it may seem that the Construction Design and Management Regulations, which apply only to the UK, are not relevant to this wider audience. The Regulations were a response to the European Union directives and an attempt to codify good practice that is relevant to all professionals in the construction field. They provide a basis for good practice which can be used as a template in any area of the world and are essential reading for any organisation offering design services who wishes to achieve ISO 9000 quality certification (see Chapter 12).

I have chosen to combine the review of the Regulations with the examination of the role of a Contractor's Design Co-ordinator to compliment the emphasis on Design and Construct contracts within the contract review.   It is the Design Co-ordinator who is most likely to deal with the Planning Supervisor and Designers, which is central to the CDM Regulations. The more commercial aspects of the Design Co-ordinator's role are taken after the review of the Regulations.

Consultation was underway at the time of publishing on a review of the Regulations with a view to implementation in October 2006. The major change trailed is a proposal to abolish the Planning Supervisor and add his responsibilities to the Employer who will have to employ a Design Co-ordinator to carry them out.

11.2   THE CDM REGULATIONS

11.2.1   Introduction

Having extolled the virtues of brevity for impact in Chapter 8 with the detail in Appendices this chapter is in more detail and may appear repetitious with lists of similar duties for the different entities. This is because it is primarily intended as a reference for those who sublet contracts to these entities.   Even so, this text does not pretend to do more than outline the steps that the entities which the Regulations identify must comply with and how this is practically likely to affect a Contractor's site management.   A full explanation of the Regulations is expertly

done in "The CDM Regulations Explained", by Raymond Joyce, who examines both the background law as well as each individual requirement.
The Regulations define the following entities:

- The Client
- The Designer
- The Planning Supervisor
- The Principal Contractor
- Contractors.

These roles are independent of the contractual arrangements. Organisations can and do carry out several roles and roles can be passed from one organisation to another. An example of the first scenario is a Design and Construct Contractor who may be Designer, Planning Supervisor and Principal Contractor. He must identify either individuals or departments which fulfil the different disciplines. The most common examples of the second scenario are the role of Planning Supervisor during a tender period passing to the successful contracting company and the passing of the role of Principal Contractor from, say, a civil engineering company to a mechanical and electrical contractor once the fit-out has commenced.

All UK Contractors and Designers have their own forms and procedures to comply with the Regulations. Clients often build infrequently, particularly in the private sector, and have to be instructed as to what is expected of them (a duty is laid on all the defined entities by the Regulations to carry this out).

The purpose of the Regulations is to define responsibilities to ensure safe construction of projects, subsequent safe operation and ultimately safe demolition. There is a danger that compliance is seen as a box ticking operation (something that the 2006 revisions are intended to tackle) that does not deliver the intended benefits. Currently, within the Building Industry, some quantity surveying practices offer Planning Supervisor services without employing any designers, engineers or construction experts. They remain experts in the meaning and application of the Regulations which, in my personal opinion, is not in accordance with the original intention.

Having derided box ticking and checklists this outline of the Regulations, of necessity, is a series of checklists for the busy practitioner who finds himself in a position that requires him to suddenly be aware of the duties upon him. It is impossible to paraphrase the full implications of the Regulations in a single chapter. A reader may find himself first referred to this chapter because he is involved in letting a subcontract and has to be aware of the necessity to check competences.

Before reviewing the entities created by the Regulations it is sensible to understand the two sets of documentation that are central to the Regulations. The first of these is the Health and Safety Plan which forms the centrepiece of administration during construction of the project.

11.2.2   Health and Safety Plan

The Health and Safety Plan serves two different purposes.

Firstly, during the pre-construction phase of a project the Health and Safety Plan brings together health and safety information obtained from the Client and the Designers and where appropriate the Planning Supervisor.

The Health and Safety Plan must be provided to any Contractor before arrangements are made for the Contractor to carry out or manage any works. In practical terms the Health and Safety Plan should be sufficiently developed for it to form part of the tender documentation. This will enable prospective Principal Contractors to be made aware of the project's health and safety requirements. The section titles for the plan are:

1. Nature of Project
2. The Existing Environment
3. Existing Drawings
4. The Design
5. Construction Materials
6. Site-wide Elements
7. Overlap with Client's Undertaking
8. Site Rules
9. Continuing Liaison.

The second function is during the construction phase of the project when it becomes the central document within which all the potential risks and hazards of each phase of construction are analysed progressively as the project progresses. The Health and Safety Plan will draw on:

(a) the Principal Contractor's health and safety policy and assessments;
(b) the Health and Safety Plan prepared by the Planning Supervisor during the pre-construction phase; and
(c) the details on the management and prevention of health and safety risks created by Contractors and Subcontractors.

The Health and Safety Plan is a document which has to be kept under review and modified to anticipate and reflect changing circumstances (typically the shift into the fit-out stage with different trades) and the different risks encountered as the construction work progresses.

During the construction phase the Health and Safety Plan is under the jurisdiction of the Principal Contractor and, as previously detailed, this can change during the course of the project. Once the project is complete the function of the Health and Safety Plan is complete.

The second major document introduced by the CDM Regulations is the Health and Safety File which stays with the project for its operational life.

11.2.3   Health and Safety File

11.2.3.1 Introduction

The Health and Safety File at its simplest is a maintenance manual enlarged to alert those who will be responsible for a structure after handover to risks that must be managed when the structure and associated plant is maintained, repaired, renovated or demolished.

The Planning Supervisor has to ensure the File is prepared and that it is handed to the Client at the completion of the works. The Client then has to store the File for the life of the structure(s) constructed in the project.

11.2.3.2 Contents of Health and Safety File

Information contained in the File should include that which will assist persons carrying out construction work on the structure at any time after completion of the current project. This should cover maintenance, repairs, alterations, additions and demolition. The information may include:

- record drawings and plans used and produced throughout the construction process along with the design criteria;
- general details of the construction methods and materials used;
- details of the structure's equipment and maintenance facilities;
- maintenance procedures and requirements for the structure;
- manuals produced by specialist contractors and suppliers which outline operating and maintenance procedures and schedules for plant and equipment installed as part of the structure;
- details on the location and nature of utilities and services, including emergency and fire fighting systems;
- any other relevant information contained in the Health and Safety Plan.

11.3   THE DUTIES AND REQUIRED COMPENTENCES OF THE ENTITIES

The review of the duties of the entities takes three parts, which are the duties themselves, how they are discharged and the assessment of competence.

The latter is one of the harder areas and the forthcoming review of the Regulations is going to consider how current competence can be measured rather than past competence, which will be seen to be the prime item in the reviews below. Assessment of competence is included against each entity. The reader that is involved in appointing or letting a contract to that particular entity should study the requirements against it.

It is essential that a Project Manager has specific knowledge of the roles and duties of all the entities because his company is likely to have in a Design & Construct contract, the Designer, Planning Supervisor and Principal Contractor's roles and be involved in their appointment. He therefore also needs to be aware of the Client's duties in this respect as well because he will appoint them and will need to be aware of what information the Client must supply.

To avoid repeating duties the review of the Principal Contractor and Contractors is combined.

11.3.1   The Client

The Duties

1. The Client, Developer, etc. (or an agent acting on behalf of) must make a declaration with regard to who acts as the Client for the purposes of these Regulations. This declaration should be submitted with the notification of the project to the HSE.
2. The Client must appoint a Planning Supervisor which should be as soon as is practicable after there is a commitment to the project by the Client. It is the Planning Supervisor who has to ensure that the notification of the project is given to the HSE. The Client must satisfy himself that the Planning Supervisor is competent, and adequate resources will be allocated to the project.
3. The Client must ensure that the Planning Supervisor is provided with all relevant and available information.
4. The Client shall not arrange for a Designer to prepare a design unless he is satisfied that the Designer has the competence and has or will allocate adequate resources to the project.
5. The Client must appoint a Principal Contractor which should be as soon as is practicable after the Client has such information about the construction involved as will enable the Client to be able to satisfy himself with regard to the Principal Contractor's competence and resource levels.
6. The Client has to judge if the Health and Safety Plan has been developed sufficiently by the Principal Contractor prior to construction being allowed to commence.
7. The Client must ensure that the Health and Safety File is available for inspection for the life of the structure.

In this instance, the ultimate Client has to assess his own competence to decide on the competence of others. If he is a first time Client he may well wish to appoint an individual or organisation to act as the Client for the purposes of the Regulations. This body can, and often is, totally separate from the contractual running of the commercial operation of the contract and appointment under the Regulations has no impact on commercial operation under any of the contracts reviewed in this book.

Discharge of Duties

Other than making the necessary appointments and carrying out the assessment of competence (which is reviewed against each entity as a contracting organisation may also carry out similar appointments), the duties are carried out as follows.

The Client must ensure that the Planning Supervisor is provided with all relevant and available information. The information required to be provided about the state of the condition of the premises where construction work is intended to be carried out, is that which is relevant to the functions of the Planning Supervisor under these Regulations and which the Client already has or could ascertain by making enquiries which is reasonable for a person in his position to make.

The information provided by the Client will form the starting point for the Health and Safety Plan which the Planning Supervisor must ensure is produced. Some of the information may be required for the Safety File. (The information may also be required to enable Designers to carry out their work and comply with their duties under the Regulations.)

Procedure

1. Contact Service Authorities (Electric/Gas/Telephone and Water) to obtain plans of their services on and adjacent to the proposed construction site.
2. Contact Local Authorities (District/Metropolitan/County to obtain information and plans of:
   (a) any other services
   (b) any planning restrictions which might affect health and safety
   (c) any existing traffic, noise or working restrictions
   (d) previous use(s) of the site or adjacent land
   (f) any known proposed use(s) or planning applications on adjacent land
   (g) rights of way on or adjacent to the site.
3. Contact local Water/Harbour/Conservancy Authorities to obtain information on any open/tidal waters on or adjacent to the proposed construction site or any other relevant information or restrictions.
4. In urban situations contact local Library/Museum for old plans/information on past use of site and adjacent land.
5. Any information/plans which the Client has in his possession from either the purchase or the existing use of the site which are relevant to health and safety. This is particularly important where an existing plant or facility is being upgraded.

6. Contact owners and occupiers of adjacent properties for any relevant information, e.g. private services, past uses, etc.

This information should be obtained prior to arranging the site survey and soil investigation so that any matters arising can be dealt with adequately, particularly with regard to services, previous structures, contaminated land and hazardous existing materials, mining subsidence, ground conditions and structural stability.

All operational systems on a site that are being retained, included or adapted in the project must also be surveyed and their operational functions described where no adequate existing data is available (see 5 above).

Information Supplied by Client - Checklist

1. Service Authorities: Electric, Gas, Telephone, Cable, Water
2. Other Services: Oil or Fuel pipelines, Surface Water Sewers and Sewerage Watercourses/Tidal Waters
3. Site Access: existing including any temporary/permanent rights of way
4. Restrictions: planning, traffic, noise and working hours
5. Existing use: information and plans
6. Previous use(s): information and plans
7. Site Soil Investigation: Topographical survey, Geological survey, Mining subsidence, Boreholes/trial holes
8. Site Survey: Existing land, buildings, ruins etc., water courses, visible service equipment access or egress, boundaries, use of adjacent land footpaths, tracks and other routes. Full details of plant systems and operations if information in 5 above is inadequate or incomplete
9. Planned use of the new facility both current and future. (This should include adaptions, extensions and possible changes of use where relevant.)

11.3.2 Planning Supervisor

Duties and Responsibilities

1. To ensure that notice of the project to which he is appointed is given to the HSE, unless there are reasonable grounds for believing that the project is not notifiable.
2. To ensure that among the Designer's considerations adequate regard has been given to:
   (a) avoid foreseeable risks
   (b) combat risks at source, with regard to health and safety of any person
   (c) give priority to measures which will protect people generally over those that only protect the person carrying out the work

3. To ensure that the design includes adequate information with regard to any part of the project, a structure or material which might affect the health and safety of any person.
4. To ensure co-operation between Designers.
5. To be in a position to give adequate advice to the Client or any Contractor with a view to enabling them to comply with Regulations 8(2) and 9(2) regarding the competence and resources of a Designer.
6. To be in a position to give adequate advice to the Client regarding the competence of the Principal Contractor.
7. To ensure that a Health and Safety Plan is prepared and provided to any Contractor before arrangements are made for the Contractor to start construction works.
8. To ensure that a Health and Safety File is prepared, that it is reviewed, amended or added to prior to being given to the Client, and that on completion of any structure, the Safety File for that structure is given to the Client.

Discharge of Planning Supervisor Duties

The Planning Supervisor must ensure that the relevant local HSE area office is notified of the project for which he is appointed unless he has reasonable grounds for believing that the project is not notifiable.

The Planning Supervisor has to give consideration of design with regard to reducing risks to the health and safety of people. This should be done in two stages.

1. At an early stage of the design development he should examine the Designer's basic procedures and systems for dealing with these matters.
2. As the design of each structure or part of the works is completed he should, by examination of the Designer's plans, drawings, risk assessments, and by using the Planning Supervisor's knowledge and practical experience of site construction procedures, check for any omissions or anomalies in the Designer's considerations.

Any matters arising should be referred back to the Designer for further consideration. The Planning Supervisor role is not to instruct or direct the Designer in his work, but to influence and assist the Designer to produce a design which takes due account of the health and safety aspects, as required by Regulation 13(a).

The Planning Supervisor should not accept any Designer's work where he is not satisfied with regards to any health and safety matter. He should immediately inform the Client in writing if this situation occurs. If at any time during the project the works are changed or varied then the process will have to be repeated.

It is the Planning Supervisor's role to see that the Designer includes adequate information about any aspect of the project or structure or materials (including articles or substances) which might affect the health and safety of any person. By examination of the Designer's risk assessments, plans and drawings and details of any materials that the Designer has specified the Planning Supervisor should be able to check that information supplied by the Designer:

1. covers all the matters arising; and
2. that the information on each item is adequate.

Any shortfall in items or information should be referred back to the Designer for action. The Client should be informed immediately in writing if the Planning Supervisor is not satisfied with the information provided by the Designer. When the Planning Supervisor is reasonably satisfied that the information produced by the Designer meets the requirements of Regulation 13(b) then it can be incorporated into:

- the Health and Safety Plan;
- the Health and Safety File.

If at any stage of the project any matters or substances are changed or varied then the process will have to be repeated. If the proposed changes or variations are acceptable then the revised information will have to be incorporated into the Safety Plan and the Safety File.

Where there is more than one Designer engaged in the design of the structures comprising the project then the Planning Supervisor must:

- arrange a suitable means of communication of information between the Designers;
- arrange for the Designers to co-ordinate their works;
- assess how different aspects of the design interact and affect health and safety.

The above duties for the Planning Supervisor must also be performed where a Specialist Contractor is engaged on a project and his work includes some design, whether for permanent or temporary works.

Competence Assessment of the Planning Supervisor

The Client or any person during the life of the project who appoints a Planning Supervisor should make enquiries to check that the proposed Planning Supervisor has the necessary knowledge and ability to fulfil the responsibilities set out in Regulations 14 and 15(1). The level of enquiries should be tailored to match the scale and complexity of the project. The main headings which should be considered are the Planning Supervisor's:

1. knowledge of construction practice;
2. familiarity and knowledge of the design function;
3. knowledge of health and safety issues, particularly in preparing a Health and Safety Plan;
4. ability to work with and co-ordinate the activities of different Designers;
5. ability to be a bridge between the design function and construction work on site;
6. management systems to monitor the quantity and quality of the resources to be employed on the project;
7. management system to monitor the quantity of the resources employed and the quality of the results;
8. allocation of time for the different duties;
9. technical facilities available to aid the staff.

Planning Supervisors normally produce a standard document to demonstrate their general competence and to define the level of their available resources and management systems. However, complex or special projects will have to be dealt with on an individual project basis.

11.3.3 Designers

Duties and Responsibilities

1. Designing for health and safety: designers must consider the need to design in a way which avoids risks to health and safety or reduces these risks as far as practicable so that projects they design can be constructed safely.
2. Specified substances or equipment: the Designer will need to analyse the hazards and risks arising from the use of any specified products.
3. Design for future maintenance: the Designer should take into account the future need to maintain and repair the structure safely.
4. Special requirements for construction: if there are special requirements resulting from the design for the construction of a particular part of the works then the Designer must give sufficient information to enable the works to be carried out safely.
5. Future works: designers have to inform the Planning Supervisor of design features relevant to future work on the completed structure to enable the Planning Supervisor to produce a Health and Safety File.
6. Co-operation: designers are required to co-operate with the Planning Supervisor and well as with other designers.
7. Principles of prevention and protection to be used throughout the design.

Discharge of Designer's Duties

Designing for health and safety: the duty imposed on designers by Regulation 13 is qualified by the phrase "so far as is reasonable practicable". In determining what is "so far as is practicable" the risk to health and safety produced by a feature of the design had to be weighed against the cost of excluding that feature by:

- designing to avoid the risks to health and safety;
- tackling the causes of risks at source; or if this is not possible;
- reducing and controlling the effect of risks by means aimed at protecting anyone at work who might be affected by the risks and so yielding the greatest benefit.

The cost is counted not just in financial terms but also those of fitness for purpose, aesthetics, buildability or environmental impact. The overall design process should not be dominated by the need to avoid all risks during the construction phase and maintenance. But by the application of these principles it may be possible to make decisions at the design stage which will avoid or reduce risks during construction work and so make it easier for contractors to devise safe systems of work.

An integral part of the design function, when decisions are being taken regarding the balance between design considerations and the effect upon the health and safety of construction workers, is the need for the Designers to carry out risk assessment.

As the design develops, the Designer should examine methods by which the structure might be built and analyse the hazards and risk associated with these methods in the context of his design choices. The need to avoid risks completely or to tackle them at source or by reducing or controlling their effects should then be positively applied to the hazards, and the design decisions rejected, altered or confirmed as appropriate.

A designer specifying substances or equipment for use during the construction work will need to analyse the hazards and risks arising from their use and, where possible, avoid or reduce risks by appropriate selection.

If the Designer cannot avoid the use of substances or equipment which he has identified as having a hazard or risk during the construction works then this information must be passed on to the Planning Supervisor for incorporation into the Health and Safety Plan. The information the Designer should be giving to the Planning Supervisor includes:

- the name of the product;
- the name and address of the manufacturer;
- the manufacturer's instructions for the product's use;
- the perceived hazard or risk;
- the significance of the hazard or risk;
- the reason that the use cannot be avoided.

The Designer should consider the future need to maintain and repair the structure safely and give particular attention, in preparing the design, to avoiding or reducing risks when selecting safe means of access for those activities. The Designer should consider the information from his design that would be required for the safe carrying out of the future maintenance and repair works that could reasonably be foreseen. This information should be submitted to the Planning Supervisor for inclusion in the Health and Safety File.

A design may be such that it requires particular attention by the Contractor when considering the method of construction. The Designer should make clear the principles of the design and describe any special requirements for the purposes of construction.

Where a structure has, as part of its design, anything which could foreseeably prove to be a hazard in the course of any later construction work, including maintenance, repair, alteration or demolition of the completed structure, appropriate information should be included with the design passed to the Planning Supervisor for inclusion in the Health and Safety File.

Designers are required to co-operate with the Planning Supervisor and with any other Designer who is preparing any design in connection with the same project or structure so far as is necessary to enable each of them to comply with the requirements and prohibitions placed on him in relation to the project by or under the relevant statutory provisions.

The health and safety aspects of design should be communicated to, and where necessary discussed with, the Planning Supervisor so as to:

- avoid or reduce risks arising from any interaction with the work of others involved in design and planning;
- enable the information to be incorporated into the Health and Safety Plan.

The Planning Supervisor may suggest changes arising out of the consideration of the overall design and designers should co-operate to enable both to comply with their respective duties.

The principles of prevention and protection to be considered by the Designer are as follows:

- If possible avoid the risk completely, by using alternative methods or materials;
- Combat risks at source, rather than by measures which leave the risk in place but attempt to prevent contact with the risk;
- Wherever possible adapt work to the individual particularly in the choice of work equipment and methods of work. This will make work less monotonous and improve concentration, and reduce the temptation to improvise equipment and methods;

- Take advantage of technological progress, which often offers opportunities for safer and more efficient working methods;
- Incorporate prevention measures in a coherent plan to reduce progressively those risks which cannot altogether be avoided and which take into account working conditions, organisational factors, the working environment and social factors. On individual projects, the Health and Safety Plan will act as a focus for bringing together and co-ordinating the individual policies of everyone involved. Where an employer is required under Section 2(3) of the Health and Safety at Work Act 1974 to have a Health and Safety Policy, this should be prepared and applied by reference to these principles;
- Give priority to those measures which protect the whole workforce or activity and so yield the greatest benefit, i.e. give collective protective measures, such as suitable working platforms with edge protection, priority over individual measures, such as safety harnesses;
- Employees and self-employed need to understand what they need to do which can be accomplished by training, instruction and communication of plans and risk assessments;
- The existence of an active safety culture affecting the organisations responsible for developing and executing the project needs to be assured.

Competence of Designers

The Client (nor any other person thereafter if the responsibility for design does not remain with one organisation) shall not arrange for a designer to prepare a design unless he is satisfied the Designer:

1. has the competence for; and
2. has or will allocate adequate resources to, the project.

The Client will need to check the knowledge, ability and resources of the Designer to carry out the duties set out in Regulation 13. Those who engage designers should seek advice where necessary, and the Planning Supervisor must be prepared to give adequate advice as required by Regulation 14(c). The main headings to be considered are the Designer's:

1. knowledge of the construction process for the project;
2. appreciation of the impact of design on health and safety;
3. awareness of relevant health and safety legislation;
4. use of appropriate risk assessment methods;

5. health and safety policies as an employer and for design work carried out;
6. resources employed on the project with regard to skill and training;
7. system for reviewing the design against the requirements of Regulation 13;
8. allocation of resources for the various elements of the Designer's work;
9. technical facilities to support the Designers;
10. method of communicating design decisions to ensure that the resources to be allocated are clear;
11. method for communicating the remaining risks after the duties in Regulation 13(a) have been complied with.

Designers normally produce a standard document to demonstrate their general competence and to define the level of their available resources and management systems.

Whereas this assists the Client in his duties in Regulation 8(2) and 9(2) it is likely that complex or special projects will need to be dealt with on an individual basis.

11.3.4   Principal Contractor and Other Contractors

The only significant differences between the duties and responsibilities of a Principal Contractor and a Contractor are that the former is responsible for and is "keeper" of the Health and Safety Plan and that the latter communicates through the Principal Contractor. This single section will therefore suffice for a Project Manager to determine his own responsibilities and to provide a checklist for anyone letting a subcontract (all subcontractors are classed as Contractors under the Regulations).

The Principal Contractor's Responsibilities and Duties

1. To develop the Health and Safety Plan so that it has the relevant features with regard to health and safety for the actual work being undertaken.
2. To examine the Health and Safety Plan and the risk assessments of other Contractors to confirm that the seriousness of the risks have been properly evaluated.
3. To ensure co-operation between all Contractors on the site.
4. To ensure that all Contractors and every employee working on the site comply with any rules contained in the Health and Safety Plan.
5. To take reasonable steps to ensure that only authorised persons are allowed on site.
6. Ensure that any notice given under Regulation 7 is displayed.

7. Promptly provide the Planning Supervisor with information.
8. To ensure that information on the risks to health and safety is provided to each Contractor and employee.
9. To ensure arrangements exist for persons who work on site to be able to offer views and give advice on health and safety issues.
10. To co-ordinate any shared use of equipment.

Discharge of Principal Contractor's Duties

The Health and Safety Plan, which the Planning Supervisor must ensure is prepared, has to be developed by the Principal Contractor. The Health and Safety Plan is the foundation upon which the health and safety management of the construction phase should be based. Responsibility for the Health and Safety Plan should be transferred from the Planning Supervisor to the Principal Contractor as soon as possible after this appointment is made to allow the maximum time for further development of the Health and Safety Plan before construction work commences. The Principal Contractor should develop the Health and Safety Plan so that it:

- incorporates the approach to be adopted for managing health and safety by everyone involved in the construction phase;
- includes the risk assessments prepared by Contractors under the Management of Health and Safety at Work Regulations 1992 and other legislation;
- incorporates the common arrangements (including welfare); these may be imposed by the Client or developed by the Principal Contractor;
- includes arrangements for fulfilling the Principal Contractor's duties under Regulations 16 to 18;
- includes reasonable arrangements for monitoring compliance with health and safety law;
- includes, where appropriate, rules for the management of the work for health and safety;
- includes rules on the shared use of equipment;
- can be modified as work proceeds according to experience and information received from the contractors.

Regulation 3 of the Management of Health and Safety at Work Regulations 1992 requires contractors to prepare risk assessments which should address risks to employees and to any other person who may be affected. Three aspects of the risk assessments prepared by the Contractors on a project will particularly influence the role of the Principal Contractor:

1. the seriousness of the risk

2. the nature of the assessment
    Certain activities of a Contractor:
    a) will remain the same from project to project, and the same risk assessment will be sufficient and suitable for all projects;
    b) will vary from project to project and the risk assessment may have to be modified;
    c) will change so much that a fresh risk assessment will be required.
3. the inter-relationship with other assessments. Some activities of a Contractor:
    a) will have no effect on other contractors working on the same project;
    b) will in certain circumstances, or at certain times, affect other contractors;
    c) will affect other contractors.

The Principal Contractor should examine the Health and Safety Plan and the assessments of other contractors to confirm that the seriousness of the risks have been properly evaluated, to ensure that the assessments have where necessary been adapted to or prepared for the project, and to identify those where an inter-relationship problem might exist.

In certain circumstance on a construction project, a number of contractors may be exposed to the same risk. It may be appropriate for the Principal Contractor to co-ordinate the preparation of a single risk assessment common to all the contractors concerned.

Where an interaction problem exists or when a common assessment has been prepared, the Principal Contractor should take a positive role in ensuring that the general principles of prevention and protection are applied when deciding on the measures which need to be taken as a result of the assessment, and on the arrangements required under Regulation 4 of the Management of Health and Safety at Work Regulations 1992.

The agreed measures and arrangements:

- must deal with the risk in a co-ordinated and effective way;
- may involve appropriate action to all Contractors involved or to one Contractor acting on behalf of the others; and
- should normally be incorporated into the Health and Safety Plan.

The Principal Contractor shall take reasonable steps to ensure co-operation between all contractors sharing the construction site so far as is necessary to enable each of those contractors to comply with the requirements and prohibitions of the relevant statutory provisions.

Regulation 9 of the Management of Health and Safety at Work Regulations 1992 requires employers and the self-employed sharing a workplace to co-ordinate their activities, co-operate with each other and share information to help each to comply with their duties under the relevant statutory provisions.

The other Management of Health and Safety at Work Regulations will require varying degrees of co-ordination and co-operation for their effective implementation. Individual contractors may be able to fulfil some of the requirements but where there is interaction there must be a management system for its control.

These management arrangements (see also Chapter 12) must be spelt out in the Health and Safety Plan. The Principal Contractor will normally have to take a leading role in preparing co-ordinated emergency procedures for the project as a whole, in establishing co-operation between Contractors and in ensuring that relevant information is exchanged.

Where the construction work is on the Client's operating premises the Client may wish to take the leading role for the emergency arrangements. The role of the Principal Contractor under Regulation 16(1)(a) is to manage and give practical effect to the duties upon individual contractors to ensure an integrated approach to health and safety on site.

The Principal Contractor is to ensure, so far as is reasonable practicable, that every contractor, and every employee at work in connection with the project, complies with the rules contained in the Health and Safety Plan. The Principal Contractor may include in the Health and Safety Plan rules for the management of the construction work which are reasonably required for the purposes of health and safety. Any rules in the Health and Safety Plan must be in writing and must be brought to the attention of persons who may be affected by them.

Authorised persons should be authorised individually or collectively by the Client or the Principal Contractor. They may be authorised to enter all or just a specified part of the site. They may include contractors or employees carrying out construction work or persons who need to enter the area for purposes connected with their work. Persons who have a statutory right to enter the area where the construction work is taking place are "authorised persons".

The Principal Contractor will need to take steps to ensure that everyone else is excluded from the work area. How this is done depends on the nature and location of the project. Special consideration will need to be made for rights of way across the site and where the site forms part of or is adjacent to other work areas.

A copy of the Notice which the Planning Supervisor has to ensure is sent to the HSE (Regulation 7) should be clearly displayed on site by the Principal Contractor. The notice should be conveniently sited so that it can be read by those working on or affected by the site. Depending on the nature of the site, positioning could be at:

- the site entrance;
- a prominent place on the perimeter;
- the site office; or

- for internal work at the entrance of the area where work is taking place.

On larger sites, it may be necessary to display the notice at several locations. The method of display will depend upon the position, and must take account of likely exposure to the weather, mud and dust. The notice should be protected, and the display cleaned or renewed as necessary, so that the notice remains legible. The Principal Contractor should ensure that all contractors are aware of the contents of the Notice to enable them to comply with their duties under Regulation 19(2)-(4).

The Principal Contractor shall promptly provide the Planning Supervisor with any information which:

- is in the possession of the Principal Contractor or which he could ascertain by making reasonable enquiries of a Contractor;
- it is reasonable to believe the Planning Supervisor would need to include in the Health and Safety File;
- is not in the possession of the Planning Supervisor.

The Principal Contractor shall ensure, so far as is reasonably practicable, that:

1. Every Contractor is provided with comprehensive information on the risks to health and safety of that Contractor or any employees of persons under the control of that Contractor, arising from the construction work.
2. Every Contractor who is an employer provides any of his employees carrying out the construction work:
   (a) any information which the employer is required to provide those employees in respect of that work by virtue of Regulation 8 of the Management of Health and Safety at Work Regulations 1992;
   (b) any health and safety training which the Employer is required to provide those employees in respect of that work by virtue of Regulation 11(2)(b) of the Management of Health and Safety at Work Regulations 1992.

The information which the Principal Contractor is required to provide to contractors should be incorporated in the Health and Safety Plan. The Health and Safety Plan should be made available to contractors when they tender. This will enable the contractors to make full provision in their tenders. The Health and Safety Plan should be reviewed by the Principal Contractor to ensure the information it contains will be comprehensible to the contractors and provide them with the relevant parts.

The Health and Safety Plan should also form the basis for the provision of information by employers to their employees about the emergency procedures required under Regulation 7 of the Management of Health and Safety at Work Regulations 1992.

Similarly, the arrangements for the provision of training by employers should be covered in the Health and Safety Plan. In both cases the Health and Safety Plan should include arrangements for monitoring compliance.

The Principal Contractor shall:

- ensure that employees and self-employed persons at work on the construction work are able to discuss, and offer advice to him on, matters connected with the project which it can be foreseen will affect their health and safety;
- ensure that there are arrangements for the co-ordination of the views of employees on the construction work, or of their representatives, for reasons of health and safety.

Where safety representatives have not been appointed or do not provide complete coverage, the Principal Contractor should make arrangements, tailored to the size and nature of the project, so as to enable co-ordination of the views of employees working for different contractors to be taken into account.

Regulation 4 of the Provision and Use of Work Equipment Regulations 1992 place duties upon individual contractors to ensure that work equipment is safe and that it is used safely.

Where work equipment is shared by a number of contractors, its provision and use should be co-ordinated by the Principal Contractor through the mechanism of Regulation 9 of the Management of Health and Safety at Work Regulations 1992.

Depending on the type of equipment, the nature of the project and the contractual arrangements, the Principal Contractor may decide to take the responsibility as co-ordinator for achieving compliance with the Regulations on behalf of all common users, or may direct another contractor or group of contractors to do so.

Co-operation and exchanging information is vital in such circumstances to ensure that faults or changes in the conditions are reported to the Co-ordinator for the equipment and those instructions or limitations on use are passed to the common users. Details regarding control, co-ordination and management of shared equipment should be specified in the Health and Safety Plan.

Competence Assessment of Principal and Other Contractors

The Client shall not appoint as Principal Contractor any person who is not a Contractor as defined in Regulation 2(1). No person shall arrange for a Contractor to carry out or manage construction works unless he is satisfied that the Contractor has:

- the competence; and
- has or will allocate adequate resources to the project to comply with the requirements and prohibitions imposed by the relevant statutory provisions.

The Principal Contractor must be given sufficient time to adequately develop the Health and Safety Plan before the construction phase can commence. Those who engage the Principal Contractor should seek advice where necessary, and the Planning Supervisor must be prepared to give adequate advice as required by Regulation 14(c) as to the proposed Principal Contractor's competence.

The matters to be considered regarding the Principal Contractor's competence by the Client and the Contractor's (normally subcontractors) competence by the Principal Contractor (normally the Main Contractor) are:

1. people to carry out or manage the work, their skills, knowledge, experience and training;
2. time allowances to complete the various stages of the construction work without risks to health and safety;
3. proposals for employing people to ensure compliance with health and safety law;
4. technical and management systems for dealing with the risks specified in the Health and Safety Plan.

Principal Contractors normally produce a standard document to demonstrate their general competence and to define the standards of their management, systems and resources. Clients and Main Contractors normally include in any pre-selection procedure the need for such a document as a criterion prior to issuing any invitation to tender. The outstanding information specific to the project should be required to be submitted with the Contractor's tender.

11.4   COMMERCIAL DESIGN CO-ORDINATION

11.4.1   Introduction

All forms of Design and Construct contracts have as their starting point a set of Employer's Requirements which detail what is to be constructed, where and when. The freedom for the Design and Construct Contractor varies enormously and with it the role of the Contractor's Designer Co-ordinator. At one end of the scale a Client can have a site on which he wants a facility which he describes in terms of square metres for particular functions and the way they interact, which enables very different approaches to be taken. (I have even specified a particular ambiance for rooms in a Health Hydro.)

At the other end of the scale, the Client may have a fully designed scheme and wishes to novate the Designer's contract to the Design and Construct Contractor to avoid the risk of the frequent delays during construction that revolve around design issues. (Even is this situation the Contractor has to satisfy himself as to the Designer's competence as regards the Health and Safety criteria.)

An in-between stage is one increasingly common in Highway contracts in the UK where there is an 'Indicative Design' for which no calculations are provided and the Design and Construct Contractor has to undertake specific site investigation for individual structures and also detailed design of all structures. The finished project is usually indistinguishable from the original general arrangement drawings.

The type of Designer for the various options varies but in addition to the CDM checklist requirements the Design Co-ordinator will be looking for the following if there is not a long-term partnering arrangement with a Designer:

1. experience of the particular project;
2. current capacity (no-one wants a past record built on staff who have since left);
3. standing in the industry: if part of the tender team will the Designer's name be an asset?
4. current Professional Indemnity Policy;
5. ability to meet the design programme and willingness to accept contract conditions which make the Designer liable for the Contractor's costs of delay in providing the design information, as well as for incorrect information;
6. the Designer's role, if any, in the quality assurance aspect of the contract. Highways Agency contracts require the Contractor's Designer as an entity to sign off the works;
7. ability to work within the company's computer-aided design (CAD) (see below).

The letting of a design contract follows the same procedure as that for a subcontractor.

During the design process the Design Co-ordinator will seek with the Designer to make savings within the design by researching different materials and techniques which are either cheaper in themselves or allow quicker construction. Whether or not the savings are shared with the Client depends on the form of contract. This process must not impinge upon delivery of the design.

In most cases the Design Co-ordinator will be central to managing the quality assurance packages that will, or should, form a major part of the contract administration as there is no third party overseeing the construction in a Design and Construct contract.

It is not unusual for him to be responsible for assembling the Health and Safety File, which has been discussed in detail above.

11.4.2   Computer Aids to Design Control

Nearly all commercial design is now carried out using CAD techniques. The Design Co-ordinator has to manage, normally within a company's own system:

- Project layout grids;
- Design ownership of layouts at any particular time;
- Access to current layouts by different Designers;
- Feedback on loadings of design elements from particular Designers
- Zones for particular services.

A fully integrated system enables design alternatives to be inserted with a quick analysis of whether there are any other knock-on affects elsewhere; this helps achieve optimisation of the design.

The main benefit is the far greater certainty of fit between the various elements of the design.

CHAPTER 12

Quality, Environmental and Safety Management

12.1   INTRODUCTION

This chapter covers the details of what is required to comply with international standards to introduce construction managers to the intent behind the management systems.

The company the reader works for will either have a system which complies with the codes described in this chapter and be certified or not. A site construction manager is not able to install such a system himself. This book is neither intended to be an authority on the subject nor a reference work for setting up a system that complies.

Where an organisation subcontracts to a company that requires a compliant system or details of the procedures followed for a particular project, it may assist the Subcontractor in identifying and producing an acceptable statement regarding its management procedures.

Whether or not an organisation has a management system that complies with the standards described, it still has to comply with the quality required by the contract, together with Environmental and Safety Regulations and Law. (It may be that a specific contract requires that each organisation within the contractual chain have a compliant system and, in this case, it would be a precondition of tendering. Even then exceptions are still made for organisations at the bottom of the supply chain.)

The moves towards quality standards commenced during WWII to provide coherence between manufacturers supplying to the War effort. These, unlike construction, did not have third party inspection of all activities by Clerks of Works. Construction continued longer with its own system based solely on third party inspection until it, too, embraced quality systems, driven by other factors including:

- the rise of Design and Construct with effective self certification;
- the need to reduce costs of inspection;
- to provide error free right first time production;
- to provide a framework for continual improvement.

Internationally, this is a growing field where ever greater numbers achieve registration for both ISO 9000 for quality management and ISO 14001 for environmental protection.

This review recognises that there is a move towards integrated management systems which can be audited to accord with both ISO 9000 and 14001.

The field of safety management in the UK is lead by OHSAS 18001:1999 which is based on ISO 14001, but is, at the time of publication, largely an unaccredited scheme. That does not mean that it cannot form part of an integrated scheme.

For some readers the overview will be enough before they read the details of their own organisation or company's procedures. Others may prefer greater detail provided thereafter, particularly if they are drafting a statement of procedures as a subcontractor. The safety procedures in the final element are relevant to all and there is some duplication of what is necessary to comply with CDM Regulations considered in Chapter 11.

12.2   AN OVERVIEW

ISO 9000 and ISO 14000 are families of standards referred to under these generic titles. They consist of standards and guidelines relating to management systems and related supporting standards on terminology and specific requirements such as auditing.

In ISO 9000 the standardised definition of quality refers to all features of the service or product which are required by the Customer, Client or Employer. Quality Management is what the company or organisation does to ensure that its service or products satisfy the Employer's quality requirements and comply with any regulations applicable to those services or products.

ISO 14000 is concerned with what the company or organisation does to minimise harmful effects on the environment caused by its activities.

Both ISO 9000 and ISO 14000 concern the process or way in which the organisation goes about its work and not the end result.

Thus, in the former case, it is intended to demonstrate that everything has been done to ensure that the product or service satisfies the Employer's quality requirement. In the latter, that everything has been done to ensure that a product will have the least harmful effect on the environment at any stage of its life cycle from construction to demolition, either by pollution or by depleting natural resources.

Central to both of these standards is external examination by an "Accredited" body; these issue certificates giving written assurance that they have audited the organisation's management systems and verified that it conforms to the specified standard.

"Registration" is achieved by the "Accredited" body recording the certification in its Register.

12.3   ISO 9000 QUALITY SYSTEMS

The ISO 9000 family of documents consists of:

- ISO 9000 Quality Management Systems fundamentals and vocabulary installation and servicing;
- ISO 9001 Quality Management Systems requirements;
- ISO 9004 Quality Management Systems guidelines for performance improvement;
- ISO 10011 Quality Systems Auditing;
- ISO 10011-1 Auditing;
- ISO 10011-2 Qualification criteria for quality systems auditors;
- ISO 10011-3 Management of audit programmes;
- ISO 19011 Guidelines on Quality and Environment Management Systems Auditing.

The original base document ISO 9000 was superseded in the main in 2000 by ISO 9001. Compared with the previous standard this:

- applies to all product categories, sectors and organisations;
- reduces the required amount of documentation;
- connects management systems to organisational processes;
- is a natural move towards improved organisational performance;
- has greater orientation towards continual improvement and customer satisfaction;
- is compatible with other management systems such as ISO 14001;
- is capable of going beyond ISO 9001:2000 in line with ISO 9004:2000 in order to further improve the performance of the organisation.

The most important change is the third of these, as it effectively changes the requirement from procedural (how activities are controlled) to process based, being a statement of what is done.

ISO 9001:2000 is simpler and follows the principles which also exist in BS EN 9000 of:

- say what you do;
- do what you say;
- check that you have; and
- act on the checks.

As every company is organised slightly differently even in the same market sector, each set of documentation and procedures will be slightly different. As construction civil and process engineering is such a wide field it is impossible to give examples of what would be relevant or included in every case.

There are five main requirements within ISO 9001:2000 and these are:

1. The Quality Management System

This will be in two parts, being:

a. Quality Manual

This high level document will describe:
- the intention to operate in a quality manner;
- the business aims and range of products/services;
- how the standard is being applied;
- how the business operates;
- provision for external audit.

b. Procedural Manual

This defines, inter alia, how:
- documentation is controlled;
- communications are controlled internally and externally;
- each activity which impinges upon the quality of the service or product is executed in the simplest terms to facilitate audit;
- control of nonconforming product or service is executed;
- procedures for subcontracted work;
- internal audits are undertaken.

2. Management Responsibility

This would typically consist of a Management Tree with the responsibility for each aspect, identified in the other four sections, defined. One senior member of staff must be designated the "Management Representative" who is the focal point for the system. Within this section is the provision for a high level management review.

3. Resource Management

This covers the resources necessary for the customer to receive the agreed service or product under the contract. In the construction industry it is almost inevitable that each team is built per project, as are the physical resources. The plan is likely to say this and how they will be assembled and the roles and responsibilities divided and communicated.

4. Product Realisation

This will describe the process from taking the order to ensuring that the works are to the specification and drawings and are completed to time.

5. Measurement Analysis and Improvement

This will cover:

- checks on the quality of the works including frequency;
- measure of employer satisfaction;
- monitoring and internal audit of systems;
- reasons for non compliance and action to eliminate it in the future;
- evaluation of subcontractors for future projects;
- improvements possible to the systems; and
- training requirements.

The system should contain provisions so it can heal and manage itself, but all this data in statistical form will be presented for the "Management Review".

12.4   ISO 14001 AND ENVIRONMENTAL MANAGEMENT SYSTEMS

Under today's legislation, implementing Environmental Management Systems (EMS) can help to reduce the risk of prosecution. ISO 14001:1996, (which superseded British Standard 7750), and the Eco-Management and Audit Scheme (EMAS) - EC Regulation 761/2001, provide guidance on establishing management systems that can help to prevent or minimise pollution risks, meet legal requirements and improve business performance. The system itself can be structured in any way and can follow that described for quality above.

The operational controls in construction are aimed, in the main, at the specific purposes of complying with legislation and controlling actual or potential impact of the products and services on the environment.

The starting point is a register of "Environment Aspects" which, basically, are emissions to air, land or water that may harm the Environment in the short, medium or long term. ISO 14001 requires that each Environment Aspect should be considered not only in terms of normal operations but also abnormal operating conditions, as well as reasonably foreseeable emergency conditions such as fire, flood, major spills, etc.

Each "Environment Aspect" should be given a significance rating, the process for which should be recorded and will probably revolve around:

- applicable legislation;
- quantity;
- scale;
- severity;
- likelihood of incident; and
- impact on the community.

These matters can be included within the Environment Policy document which will contain a commitment to continuing improvement and compliance with legislation. The policy should be publicly available and include a procedure for management review.

The remainder of the EMS should include:

- a system for receiving updates on environment law;
- objectives can be set with a programme to achieve them based on the register of Environment Aspects;
- reporting responsibilities should be clear;
- site inductions and subsequent follow-up training should be detailed using the register as a base;
- operational responsibilities and emergency procedures and responsibilities should be clearly identified;
- systems for recording internal and external communications;
- monitoring that the required performance is being achieved;
- procedures for dealing with non conformity; audits by suitably qualified personnel; and
- management review of the systems as provided for within the Quality Systems.

Within Europe there is a parallel scheme known as EMAS which grew out of European Council Directive 1836/93 which is voluntary. BS 8555:2003 was produced to provide a guide to establishing an EMS system to either standard.

12.5   HEALTH AND SAFETY MANAGEMENT

In 1996 the BSI produced BS 8800 which provided guidance on management systems for health and safety. There was a further attempt at standardisation in November, 1998, culminating in OHSAS 18001:1999 which resulted in dividing the requirements into the following sections. These systems can be applied in the UK to achieve the requirements of the CDM Regulations (see Chapter 11).

12.5.1   Policy

The organisation should clearly define its Health and Safety policy in a document and communicate it to all employees. The policy should be reviewed and amended as required. The organisation's top management should define and endorse the policy, which should be documented, implemented, relevant and maintained regularly. At the very least, the policy should cover the following subjects:

- recognition of health and safety as an integral part of the business;
- recognition that Health and Safety is a line management responsibility from the top down;

- involvement and consultation with employees;
- provision of adequate training to meet competence levels;
- compliance with legal requirements and publishing of objectives;
- provisions of adequate resources;
- Cost effective improvements in Health and Safety performance;
- Understanding and implementation at all levels;
- Reviewing, auditing and maintenance of the system.

Occupational health and safety considerations need to be planned so that success or failure can be clearly seen. This will involve clear identification of requirements, the criteria, responsibility, timescales and performance measurement.

12.5.2 Risk Assessments

The main tool for control of Health and Safety is hazard identification, risk assessment and risk control.

The Management of Health and Safety at Work Regulations 1999 require all employers and self-employed people to carry out a risk assessment of their hazards and follow up with periodic reviews of its existing system, as required. The objective of risk assessment is to see whether the current or planned arrangements are adequate in all circumstances. Areas covered could include:

- Materials
- Machinery/Equipment
- Processes
- People
- Premises.

A risk assessment should look at all activities, not only production or processes, but anywhere where hazards may occur (i.e. offices, canteens, toilets etc.). Following the risk assessment, the results should prompt the following action:

- Trivial - No further action and no record required;
- Tolerable - A risk that has been reduced to a level that can be endured by the organisation, having regard to its legal obligations and own H&S policy;
- Moderate - The risk needs to be evaluated carefully and reduced to being a 'tolerable risk'. Specialists should possibly be used to carry out assessment profiles and suggest risk reduction methods;
- Substantial - A high level of monitoring and record keeping will be required, until the risk is reduced or eliminated. Work should not take place while the risk assessment and urgent remedial work is being carried out. If the assessment relates to a new process or material, the activity should not be initiated;

- Intolerable - This level is not acceptable and all work should be ceased until the risk has been reduced to one of the above.

12.5.3   Action plans

These should consider the following points which the reader will recognise from Chapter 11:

- elimination of the risk;
- reducing the risk, if possible;
- if risks cannot be eliminated or reduced sufficiently, personal protective equipment should be issued;
- emergency procedures should be explained and practised.

12.5.4   Legal and Other Requirements

The management system should ensure that there is a method for identifying new requirements or current practices and any requirements for new processes or activities. Other requirements may originate from Approved Codes of Practice or similar publications.

12.5.5   Objectives

The organisation must establish objectives, which are measurable and relevant to the Health and Safety policy. These may also be formulated to achieve improvements in the number of working days lost from work-related injury and ill health.

12.5.6   Management Programme

The management shall detail the designated responsibility and time frames for achieving the objectives detailed in the system.

12.5.7   Implementation and Operation

Responsibilities for Health and Safety need to be defined by the most senior person in the organisation. Health and Safety should be an integral part of an individual's business activity and they should take responsibility for Health and Safety resulting from their activities. Individuals should also be aware of the influence that their actions can have on people outside their workplace.

12.5.8   Training Awareness and Competence

For the Health and Safety policy to succeed, the organisation will need to ensure that all personnel understand their individual influence and contribution to a successful health and safety policy. For some members of staff, there will be aspects of risk control or other activities that require specific competence. To achieve this, training will be necessary, as well as proof that the training has been effective so that such competence is evident.

12.5.9   Consultation and Communication

The key to achieving a high level of awareness in Health and Safety is effective communication. This requires open, effective circulation of Health and Safety information to keep all members of staff informed. Communication may also involve the provision of specialist advice and services from internal or external sources. Communication needs to operate at all levels within the organisation so that employees are both informed and consulted.

12.5.10 Documentation

Establishing a Health and Safety system is bound to create documented information. This will probably be information on hazards or activities and control methods. The specification requires that this information is available, described, maintained and provides direction to related documentation.

12.5.11 Document and Data Control

Having established the requirement to produce documentation for the Health and Safety system, companies need to control this information. When changes are made, they need to ensure that obsolete information is withdrawn, and that everyone who needs to know is made aware that there are new documents.

12.5.12 Operational Control

Health and Safety must form an integral part of an organisation's operation. This applies equally to a "sole trader" as it does to multinational companies in all commercial activity. Organisational arrangements must include the skills and ability to manage activities and meet legal requirements.

When deciding on individual responsibilities, companies also need to address the matter of authority for taking any necessary action. They should also provide adequate resources to enable personnel to carry out the Health and Safety policy. This section would also include details of any permit to work systems (i.e. hot works, confined spaces, etc).

12.5.13 Monitoring and Measurement

A key part of any Health and Safety system is measuring whether actions taken have been effective in both quantitative and qualitative terms. Measurement should be at an appropriate level, suitable for the organisation. Proactive measurement should be used to monitor compliance to systems and policy. Reactive measurement should be used to monitor accidents, near misses, complaints, ill health, incidents etc.

Accidents, Incidents, Non-conformances and Corrective/Preventive Actions Procedures should be established stating how to report and investigate accidents, incidents, non-conformances and corrective/preventive actions.

12.5.14 Records

During the course of operating a Health and Safety system, records will be generated. Some records will be necessary to demonstrate compliance with legal and other requirements.

12.5.15 Audit

In addition to Health and Safety monitoring, there will be a need for periodic auditing. This is a means of looking at the performance of the Health and Safety system in greater depth, to ensure the organisation is capable of achieving the required results.

12.5.16 Management Review

The management must review the Health and Safety system periodically, to ensure that the overall performance of the system and individual elements are operating as intended.

Periodic status reviews should consider additional internal or external factors, further Health and Safety improvement, (once original levels are achieved) and that the Health and Safety policy and objectives remain valid.

12.5.17 Certification

The certification process is the same as that of ISO 9001 or ISO 14001. Once the Certification Body receives a completed application form, an initial assessment is undertaken on site, to determine the state of the policy, procedures and work instructions.

If the documentation is satisfactory, then a date is set for the main assessment, which will assess the level of implementation. Once satisfactory, a certificate is issued. Annual surveillance ensures continued compliance.

12.5.18 Accreditation

OHSAS 18001 is still largely offered as an unaccredited scheme.

12.6   INTEGRATED MANAGEMENT SYSTEMS

12.6.1   Introduction

The continued development of management system standards into areas such as environment, health and safety, and information security has reinforced the calls for an integrated approach. An organisation has one management team. It therefore seems logical to adopt one management system that contains all the necessary methodologies and aims to meet all management obligations. An integrated management system is often the most effective way to discharge an organisation's obligations to its employees, customers and the wider community.

The key measure of "success" with regard to the implementation of an integrated management system is the effectiveness of the process that is developed within the company for continually improving the system. If this is achieved successfully, the system will provide a good return on the resources committed to developing and installing it.

12.6.2   Concept and Structure of Integrated Management

An integrated management system is simple in its concept. Typically, it seeks to:

- integrate applicable management system standards into one management document which provides effective control of activities within the organisation;
- identify the organisation's objectives and obligations including legal, regulatory, stakeholder and company policy issues;
- ensure that all specific requirements of the issues covered have been addressed at all appropriate locations;
- ensure that personnel receive specific training in system requirements;
- determine performance criteria, as applicable, to the system requirements;
- generate evidence that system requirements have been met;
- monitor and report the extent of compliance with the performance criteria;
- continually monitor changes to input requirements and ensure that these changes are reflected in the specific system requirements, where applicable;
- audit and analyse the system processes and amend them, where applicable;
- ensure that there is a process for continually improving the system and adopting lessons learned.

All the procedures necessary for certification of the system to ISO 9001 and ISO 14001, together with OHSAS 18001 can be carried out on the combined document.

Individuals will be described as having particular roles in the individual areas of management as well as environmental and safety.

Costs Appendix

DEBUGGING THE COSTS

An inordinate amount of time is spent in many companies challenging computer-based costing systems, often without understanding the method by which they are built up. Some Project Managers run their own manual system alongside to have information they understand and trust.

      None of this is necessary if it is clearly understood that monthly costs are a snapshot at one moment in time and that adjustments must be made to arrive at the interim result. Secondly, distrust good news as much as bad news, because the former is normally an inaccuracy which in correction leads to "adverse movements" which start accountants jumping.

The Analysis

It is equally as important to run a check if an unexpectedly favourable result is produced, to ensure that the gain is genuine and does not establish a false bench mark, as it is to check adverse movements.

Materials

The material cost is based on a series of accruals which are in turn based on delivery tickets listed on "Goods Received" sheets and valued at the material order rates. The following should be gone through:

- Are all the materials in the valuation listed on the "Goods Received" sheets?

- Have all relevant "Goods Received" sheets been included in the month's costs?
- When an invoice arrives is the system for eliminating accruals working properly? I have known the costs for items to be in twice because the computer operator could not always identify the accruals and added in the invoices as well.

Labour

Due to the weekly payment of Labour an accurate cost will almost certainly be available. The only item to watch for being the allocation of time and cost from personnel and gangs which are not exclusively based on site. This month's gain will be next month's loss as the Project Manager of the receiving contract seeks to rectify the error.

External Plant

The costs are based on a monthly statement from the site of the plant that they have had on site, with the appropriate hire periods.

Usually designated the Plant Returns, these are then evaluated at the order rates and an accrual inserted in the costs. The Plant Returns are also used as the basis of checking the invoices when they arrive, prior to payment. Points to look for:

- Are the Plant Returns fully complete? Do they recognise any and all overtime by operated plant, damage to plant and lost items?
- If Plant Returns are done on a weekly basis, are all returns relevant to the period included?
- Are accruals being identified and eliminated when the invoices are entered?

Internal Plant

Normally all internal plant is on a weekly basis and most companies also include an internal Plant Return. Plant departments are notorious the world over for not taking items off hire when they do not have anywhere else to place them and this is really all that should be checked, together with the allocation of on/off charges.

Subcontractors

The costs should reflect the amount certified by the site based Quantity Surveyor and it is difficult to get this element wrong.

THE VALUE IN THE COMPANY ACCOUNTS

At first sight it would appear that the Company accounts for operational matters should be a direct compilation of the Cost Value Reconciliations (CVRs) for each site. The policies in different companies vary greatly and indeed can vary from year to year within any one company. It is undesirable for several reasons for a year's trading to be reflected directly in the year's financial results. The main reasons are:

- Sharp increases in profits are read with suspicion by clients with particularly significant contracts;
- Similarly, falls in profit will result in worries as to the long term viability of the organisation, and when deciding on tender list (see Chapter 1) clients may opt for a company with better figures;
- Effect on share price and/or standing with the bank or creditors.

The Stock Exchange has issued rules for the preparation of company accounts and the document is generally known under the heading of SSAP with a number recognising the latest issue. Highlighted below are the critical items for construction company accounts:

- A. A conservative estimate only to be included for the value of authorised variations, possibly even value only taken for claims and extras when they are both certified and paid.
- B. Foreseen claims against the Contractor by both employers and subcontractors to be taken into account.
- C. Value for claims taken only when there is evidence in writing of acceptance of the claim and an indication of the sum to be awarded.

This obviously simplifies a complex subject. These are nevertheless the critical items which often lead to the note in construction company accounts which says either:

"The accounts have been prepared in accordance with the general principles of SSAP... with departures only to the extent that these result in a true and fair statement of the Company's position in the opinion of the Directors."

or:

"The Directors consider that the accounts give a true and fair view of the Company's position"

Such statements should be read with caution and checked against the figure for Work in Progress.

The more solid companies who regularly and like clockwork increase their profits year by year are doing so by releasing what in other companies may be considered as over provision. In extreme cases they will take no value above bill measure, even for authorised variations, until the final account is settled and paid. Any claim by a subcontractor, or potential claim, is either taken into cost or added as a liability after the computation of the cost. In this way they can be assured of taking two bad years without affecting their results. In these cases Work in Progress will be less than 8.33% of turnover (one month's turnover not paid at close of accounts).

At the opposite end of the spectrum, realistic values will be put on the outcome of the final account and a commercial view will be taken on the likely settlement of all subcontractor or other counter claims. In these cases Work in Progress will be far greater.

It is by these means that apparent stability is brought to what is and will remain a volatile business despite moves towards partnering and service contracts.

Subcontract Appendix

CONTENTS

CHECKLIST FOR CHOICE OF SUBCONTRACTORS

- Previous experience with the Subcontractor.
- The Subcontractor's ability to manage his resources and liaise with the Main Contractor's staff. Good relationships between parties are an essential requirement to developing a team approach to a successful project.
- Financial standing of the Subcontractor. His ability to wait to be paid in accordance with the subcontract.
- Within the UK the Subcontractor's ability to meet the CDM requirements (see Chapter 11).
- The Subcontractor's expertise which he can bring to the project. The Subcontractor's reputation and his standing with the client.
- The current commitment of the subcontract organisation.
- Their current workload with other contractors should be determined and serious consideration given to their ability to cope with the increased work. A large number of subcontractors just cannot say 'No' when it comes to taking on more work. They often pull and push their limited labour force between sites hoping the Main Contractor will not notice that they are stretched to the limit.
- The acceptability of the Subcontractor to the Client. On many contracts the Contractor is required to name his Subcontractors at the tender stage.
- The competitiveness of the Subcontractor's price. The price must be right otherwise the Subcontractor will never win any work. Price

discounts which may be applicable and the Subcontractor's response to negotiation may be an important factor.

- The contractual risk which the Main Contractor takes on the subcontract item. Subcontract operations of a "low risk" category may be let to the more "risky subcontractor" and hence the main contractor may include a lower subcontract price in his estimate. For example, a "high risk" subcontract operation could be external brickwork while site demolition work could be "low risk" in terms of quality and completion to time. Subcontract operations which are critical to the success of the contract require careful consideration.
- The ability of the subcontractor to meet quality assurance criteria as laid down by the Main Contractor or specified by the client.
- References available from the Subcontractor. These include trade and bank references. The willingness of the Subcontractor to allow previous contract work to be inspected. It is important that good relationships are established between the Main Contractor and Subcontractor as early in the planning process as possible. This is especially important where the main Contractor intends to sublet all the work in a series of packages on a particular project.

CHECKLIST FOR ENQUIRIES

- Details of Main Contract works
- Job title and location of site
- Name of employer
- Names of Architect Supervising Officer, Quantity Surveyor and other consultants including the Planning Supervisor
- General description of the works
- Tender or current Health and Safety Plan
- Subcontract works
- Relevant extracts from Bills of Quantities and specification
- Extracts from the contract Preliminaries section
- Copies of relevant drawings
- Details of where original documents may be inspected
- Time period for completion of subcontract work (if known)
- Approximate dates when subcontract work will be undertaken
- Name(s) of adjudicator (in case of dispute)
- Subcontractor's responsibility for site arrangements and facilities:

  1. Watching and lighting
  2. Storage facilities
  3. Unloading, hoisting and getting in materials
  4. Scaffolding
  5. Water and temporary electrical supplies
  6. Safety, health and welfare provisions

7. Licences and permits
8. Any additional facilities.

- Conditions of subcontract
- Form of subcontract agreement
- Period of interim payments and payment terms, including whether "pay when paid" will apply
- Discount applicable to the payments fluctuations or fixed price tender
- Other special conditions
- Particulars of the Main Contract conditions
- An extract from the Appendix to the form of contract will assist in providing a summary of the contract particulars. This should contain the following information:

1. Form of contract
2. Fluctuations provisions
3. Method of measurement
4. Main contract period and completion date
5. Defects liability period
6. Liquidated and ascertained damages
7. Period of interim certificates
8. Basis of dayworks
9. Insurance provisions
10. Deletions or amendments to standard contract clauses
11. Type of quotation required from the Subcontractor
12. Lump sum quotation
13. Schedule of rates
14. Other information
15. Date for the return of the tender
16. Person in the contractor's organisation to contact
17. Period for which the tender is to remain open for acceptance
18. Extent of the phasing of the works and number of anticipated visits to undertake the works
19. Reference to any other attendances likely to affect the Subcontractor.

ANALYSIS OF THE CECA BLUE FORM OF SUBCONTRACT FOR USE WITH THE ICE CONDITIONS

Clause 1: Definitions

The majority of this clause is concerned with the uncontentious establishment of definitions. Subclause (1)(b) eliminates standard printed conditions either in the Contractor's enquiry or on the Subcontractor's standard form.

With the widespread use of computers subcontractors are now putting their conditions in as part of the quotation. This is considered further in Chapter 5 under Section 5.3 "Letting a Subcontract".

Clause 2

Establishes in Subclause (1) that the Subcontractor is to complete and maintain the works to the Engineer's satisfaction as well as the Contractor's. Subclause (2) ensures that everything required in the way of temporary works as well as permanent works is included within the subcontract unless otherwise stated in the fourth schedule (Letting a Subcontract). Subclause (3) puts the Subcontractor in the same position as the Contractor in having to obtain approval for any further subletting.

Clause 3

This clause allied with Clauses 10 and 17 form the central operating principles of the subcontract.

Subclause (1) makes the Subcontractor deemed to have full knowledge of the Main Contract. The Subcontractor is entitled, at his own cost, to an unpriced set of the Main Contract documents, if requested. The Contractor must also give the Subcontractor a copy of the Appendix to the Form of Tender for the Main Contract, together with any modified Main Contract clauses, again only if requested. The former will give details of the liquidated damages, time for completion and any sectional completion. On major subcontracts it is to be hoped that the Subcontractor's estimators have seen all these documents at the time of tender and have included for the information contained therein.

Subclause (2) gives the Subcontractor all the duties and responsibilities of the Contractor and the all important obligation not to cause the Contractor to be in breach of the Main Contract.

Subclause (3) delivers the core of the subcontract making the Subcontractor indemnify the Contractor against any liability arising out of a breach of the subcontract provisions by him. Subclause (4) then specifically draws into this consideration any other contracts the Contractor has which relate to the Main Works as well as damages under the Main Contract.

Obviously the most common breach is not completing to time. The effects of this can be horrendous for a subcontractor. I have known steelwork subcontractors qualify the amount of the Main Contract damages applicable to a failure to complete by the Subcontractor.

They then believe that they have limited and quantified their maximum liability but the reality is otherwise. The following charges can be levied through the provisions of this clause linked with Clauses 10 and 17 from Subcontractors:

1. Liquidated Damages.
2. The Contractor's over-run on site based establishments.
3. The Contractor's loss of efficiency on his own works.
4. Increased costs on material supply orders as well as subcontracts.
5. Loss of earnings elsewhere.
6. Acceleration costs to mitigate delays.

7. Similar charges under 2 to 6 paid by the Contractor to Subcontractors.

The potential penalties as far as the Subcontractor is concerned are such that he must keep to programme, or to extended programme, and respond to notices given in the main under Clauses 10 and 17.

Action for Subcontractor's Agents

1. If not included in his subcontract documents he should request a copy of the Appendix to the Form of Tender and any varied or additional clauses to the ICE 5th Edition.
2. Most contractors in their subcontract enquiries say that a copy of the Main Contract documents and drawings are available for inspection. The Subcontractor's Agent should take up this offer if his company does not wish to pay for a full set of documents.

Clause 4: Contractor's Facilities

Subclause (1). Effectively this means that the Contractor shall not prevent the Subcontractor from using scaffolding that he has erected for purposes other than the subcontract works. But the Contractor has no obligation to leave it erected, or for its fitness for purpose, and also makes the Subcontractor jointly responsible for its safety if used.

Very seldom do subcontractors actually take the risk of the Contractor's scaffolding matching their needs and specify their requirements in their quotation together with all other required attendances. This is then subject to Subclause (2) or (3) of this clause and the caveats discussed above are not applicable. Subclause (2) relates directly to items that are incorporated in the Fourth Schedule Part 1 and are not for exclusive use of the Subcontractor. The obvious shared facility on most contracts is the welfare set-up. However, anything included in Part 1 does not have to be provided in the maintenance period. Subcontractors should note this when access is paramount.

Fitness for purpose of any facility provided by the Contractor is not guaranteed in Subclause (2) or (3). One of the clearest examples is if a subcontract were to include in the schedule for the provision of a 50 tonne telescopic crane, by the Contractor, free of charge. The capacity of these machines at extended radii varies greatly as does their physical size. Either of these factors might prevent the crane from being able to get close enough to the area of lift. If such a crane rated at 50 tonnes arrives in working order, but is not capable of performing the required task, the Subcontractor has no recourse. If the machine is physically capable but breaks down the Contractor is liable. It is my view that the caveat on liability for provision of plant or services "if such failure is due to circumstances beyond the control of the Contractor" can only refer to causes for which he is claiming under the Main Contract, weather delays or "act of God", etc. The problems within this example are avoided if the Subcontractor were to require and have included in the subcontract, a lift capacity at a particular point.

Subclause (3) varies from Subclause (2) in that the plant and facilities entered in Part 11 of the Fourth Schedule are for the exclusive use of the Subcontractor and by inference have also to be provided for maintenance works and in the maintenance period. Even if they are to be supplied free of charge the Contractor may well successfully contend that their supply for works of rectification is as a result of a breach and hence to be chargeable.

Subclause (4) provides the express terms in the subcontract for the Contractor to recover his costs through the subcontract for loss and damage to plant and facilities. He can deduct these from monies otherwise owed rather than claim from the Subcontractor's insurance policy.

Clause 5: Site Working and Access

It is through Subclause (1) that the Contractor can enforce the same disciplines on movement of plant as are imposed upon him by the Main Contract and for the Subcontractor to work the same site hours, unless otherwise agreed. Many Main Contracts now have restrictions on the movement of plant and hours of work which can be passed on through this section.

Countless arguments arise on possession of parts of the site but the position is quite clear; the Subcontractor is not entitled to have exclusive possession or control of any part of the site, or that such possession or control should be continuous. This seemingly harsh statement is mitigated by the requirement for access to be provided "as shall be necessary to execute the subcontractor works". Having laid the ground rules the subcontract expects reasonable operation by both Main Contractor and Subcontractor. The latter has to accept he must share the site with other contractors as well as the Main Contractor and they in turn have to recognise the requirements for the Subcontractor to execute the subcontract works. It should be noted that it is the Subcontractor, through Clause 3(1), who is deemed to have full knowledge of the Main Contract provisions and hence, the likely conditions in a well organised workplace. The Contractor has no such knowledge ascribed to him regarding the Subcontractor's possible requirements. It therefore behoves the Subcontractor to get any particular special requirements written into the subcontract document.

Subclause (3) gives the Contractor and Engineer similar rights of access to the subcontract works as the Engineer has to the works under the Main Contract.

Clause 6: Commencement and Completion

Subclause (1) makes it a condition precedent that the instruction to the Subcontractor to commence work on site is in writing. It is absolutely vital for the Contractor's Project Manager to do this.

A subcontractor under pressure, with too many commitments, will avoid coming to the site which does not instruct him in writing to commence, knowing that Clause 17 cannot be operated against him. The Contractor cannot rely upon minutes of meetings. The notice can even be a standard photostat for the small site in which the Project Manager can write in by hand the name of the subcontractor and the date of commencement and completion. It is better to have a short reference to the "provisions of Clause 17 being enacted" in the event of non-compliance.

The remainder of Subclause (1) introduces the duty on the Subcontractor to complete the works without delay, other than where sanctioned, and within the period for completion. Subclause (2) then gives the following exceptions whereby the Subcontractor may complete later than the period for completion:

- Circumstances leading to an extension under the Main Contract.
- Ordered variations to the subcontract to which (a) above does not apply. (Probably the domestic arrangements between Contractor and Subcontractor such as provision of temporary works etc).
- Breach of the subcontract by the Contractor.

The Subcontractor thereafter is entitled to an extension which is fair and reasonable but cannot exceed the extension obtained by the Contractor under the Main Contract.

The other vital provision quoted in full is: "it shall be a condition precedent to the Subcontractor's right to an extension of the Period for Completion that he shall have given written notice to the Contractor of the circumstances or occurrence which is delaying him within 14 days of such delay first occurring."

Subclause (3) lines up the subcontract with the Main Contract stage completions.

Subclause (4) eliminates the need for written notice for the commencement of off site works and it is solely the Subcontractor's responsibility to be ready by the time the Contractor orders commencement. Subcontractors should note this when signing the subcontract and endeavour to have sight of the Contract Programme or qualify if they need a mobilisation or lead time.

Subclause (5) places a requirement upon the Contractor to notify the Subcontractor of all extensions of time obtained under the provisions of the Main Contract which affect the subcontract. Obviously this includes all items which have been initiated by the Subcontractor.

Consider a case where the period for completion for a subcontract was the same as the time for completion in the Main Contract. If a separate event, entirely unconnected with the subcontract works, leads to the Contractor obtaining an extension, does he have to inform the Subcontractor? The point will eventually be tested in law, but I would say, no. It would, however be unreasonable for the Contractor to endeavour to recover costs or damages where none had been incurred. There are, of course, degrees of "affect" and I have chosen one where there is none.

Clause 7: Instructions and Decisions

Subclause (1) requires the Subcontractor, identically to the Contractor, to comply with all instructions from the Engineer or his Representative, whether valid or not. These, as always under this form, must be confirmed in writing by the Contractor. The procedure for recovery of properly incurred cost is then laid down.

Subclause (2) is one of the most frequently quoted clauses and gives the Contractor like powers to the Engineer, in respect of the subcontract works with similar compliance required by the Subcontractor, whether or not the Engineer has exercised these powers under the Main Contract. The only difference being that oral instructions on their own are not valid.

Clause 8: Variations

Subclause (1) defines the conditions through which the Subcontractor shall vary the subcontract works. All of which require orders in writing from the Contractor.

Subclause (2) prevents the Subcontractor acting on any order for variation received directly from the Engineer or Employer, whether in writing or not, unless and until confirmed in writing by the Contractor. The commonest case is with Nominated Subcontractors and they should beware of this provision which could cost them dearly.

Subclause (3) again rams home that unless the order has been in writing the subcontract works shall not be varied. Subclause (4) specifically provides for payment for variations as the contract proceeds.

Clause 9: Valuation of Variations

Subclauses (1) and (2) set out the same basic criteria for evaluation of variations as in the Main Contract. The important point is that in regard to what is reasonable the Main Contract award should be considered. In practice this effectively becomes a maximum.

Subclause (3) introduces the right of the Subcontractor to be present at the quantification of measured variations. This does not extend to their evaluation.

Subclause (4) ensures that the position with varied quantities lines up with the Main Contract and particularly Clauses 51(3) and 56(2).

Subclauses (5) and (6) bring in the provisions for daywork. Daywork rates only apply, as is the case under the Main Contract, if the work is specifically ordered in writing on daywork rates. Otherwise the provisions of Clause 9(2) apply.

It is to be hoped that the drafting of the subcontract clearly deals with the probable anomaly in Subclause (5) which has the Subcontractor being paid in accordance with the Daywork Schedule in the Bill of Quantities and in accordance with the rates and prices in his tender. Where payment is to be made on the Main Contract the Contractor shall provide a copy of the document to the Subcontractor.

This can be after the subcontract is in being because the sub clause is only advisory when it says that it "should" be included within Schedule 2.

Clause 10: Notices and Claims

Subclause (1) introduces the requirements for the Subcontractor to produce all returns and notices relating to the subcontract works that are required under the Main Contract. Ignorance of such requirements is ruled out.

Subclause (2) contains four vital sentences which are considered separately under headings (a) to (d):

(a) A duty is laid upon the Contractor to pursue the Subcontractor's case under the Main Contract. Caveats are included on proper notices, information and requisite assistance.

(b) On receiving such benefits the Contractor shall pass on to the Subcontractor such portion, including extension of time, as may be fair and reasonable. It will be difficult for the Contractor to do other than pass on the full benefit in time if it has been specifically and separately awarded. This is more equitable than the earlier versions which only required the Contractor to pass on financial benefits.

(c) The third sentence introduces the sufficiency of tender for the Subcontractor which is required of the Contractor under Clause 11(2) of the ICE Conditions. By making it for other than items claimable under the Main Contract the Subcontractor is deemed to have included for conditions and sequences reasonably required and as a result of the Contractor's method of work.

(d) This last sentence allows the Subcontractor to claim from the Contractor for delays caused by act or default of the Contractor. From the Subcontractor's point of view a valuable provision but certainly not a catch-all when (c) above is considered.

Subclause (3) has probably bankrupted more Subcontractors than any other of the contractual provisions. The Contractor is given the right to recover from the Subcontractor sums he would have recovered if the Subcontractor had given the proper contractual notices.

The basic provision is equitable because the Contractor must be able to assume that he has fully and not partially subcontracted the risks. How often though does a subcontractor, who has struggled, obviously lost money, and finished perhaps one week late, retrospectively produce a claim for four weeks extension of time?

The Contractor may well agree that he would have been due this had he notified properly. Under 6(2)(c),10(1) and 10(3) of the subcontract the Contractor's overhead, establishment and associated charges for the four weeks claimed might be deducted.

On large projects a bad job for the Subcontractor can become a disaster through not being aware of, or operating, the procedures. However, to exercise these rights in the UK the Contractor must follow the procedure particularised under Clause 15 below.

Clause 11: Property in Materials and Plant

This clause brings the Subcontractor in line with the Contractor under the Main Contract, particularly Clause 53. The Subcontractor should be wary of using valuable plant or equipment which is actually legally owned by the company which is in contract with the Contractor. I have seen a muck shifter demand to take his plant off site when a Main Contractor has gone bust only to be told it was now the Employer's property.

Clause 12: Indemnities

Through the provision of this clause the cross indemnities between Employer and Contractor under the Main Contract are passed on to the Subcontractor. The only mitigation being if the cause was "solely by the wrongful acts or omissions of the Contractor, his servants or agents". This is very difficult to prove (see Clause 13 below).

Clause 13: Maintenance and Defects

Subclause (1) requires the works to be "in the condition required by the Main Contract (fair wear and tear excepted) to the satisfaction of the Engineer". All this is to be done at the Subcontractor's expense unless it is caused by the "act, neglect or default of the Employer, his servants or agents... or of the Contractor, his servants or agents....".

Subclause (2) includes any sectional completion recognised under the Main Contract. This introduces all the Main Contract obligations to the Subcontractor to execute works of rectification in the maintenance period. Again there is the caveat for specific Contractor's responsibilities.

Clauses 12 and 13 acting together provide that all damage (breakages after installation, vandalism, etc.) is the responsibility of the Subcontractor. This is not onerous when the subcontract works are asphalting or earthworks, but take on a different complexion when the ICE Conditions and CECA Form are used for industrial developments. Subcontractors more used to the building forms, where the Main Contractor is liable for insurance of the works once installed, are often appalled to find they are liable for all damage which the Engineer requires rectified which may have occurred months after they have left site. The sums involved in these circumstances can be a significant proportion of the subcontract sum.

On multi-contract large industrial contracts the Contractor's Project Manager usually has a standard letter requiring the Subcontractor to repair the subcontract works and claim from his insurers or the perpetrators of the damage. To operate the caveats in these two clauses the Subcontractor has to effectively prove that the Employer, Contractor or other specific Subcontractors caused the damage. Vandalism is definitely at the Subcontractor's door. It is not sufficient for the Subcontractor to say that "damage of this type has been made by Contractor's plant" or "the Contractor and Employer control access to the site and therefore they have either neglected to prevent vandals entering or the damage has been caused by their servants or agents". The Subcontractor must provide proof of responsibility.

Where such provisions are excessively onerous it is up to the Subcontractor to qualify out of this in his quotation and agree an equitable compromise with the Contractor at the time of the letting of the subcontract. It might be along the lines of removing liability for impact or chemical damage when the Subcontractor has left site.

Clause 14: Insurances

The clause provides for the Subcontractor to insure the works as laid out in Part 1 of the Fifth Schedule with such insurance remaining in force to the end of the maintenance period.

Clause 15: Payment

Most reasonable Main Contractors have operated within these comprehensive rules within Clause 15(1). The specified date also benefits the Contractor in that he now has an enforceable means of gathering all his subcontract valuations for a set date which makes financial control and cost systems easier to operate.

The clause requires the date, or normally dates, on which the Contractor is to submit valuations under the Main Contract to be included in the First Schedule as specified dates. The Subcontractor must submit his valuation for payment not less than seven days before the specified date. The valuation must be a "valid statement" before it qualifies for processing by the Contractor.

This is sufficient sanction in itself to enforce compliance. In the event of the Subcontractor not submitting seven days before the specified date there is nothing to stop the Main Contractor from including in his valuation for the subcontract works and deferring payment for a month until the Subcontractor meets the next specified date. Main Contractors may seek also to use the seven day gap similarly, so subcontractors should project their statements to include work anticipated to be done by the specified date.

To be a "valid statement" the valuation must be presented in writing and "in such form and contain such details as the Contractor may reasonably require" of the value of work properly done under the subcontract, of all materials delivered to site for incorporation and, if allowable under the Main Contract, the value of off site materials.

The Contractor must then pay through the provisions of Subclause (3)(a) within 35 days. He may withhold or defer payment under the provision of Subclause 3(b) for amounts properly justified under Clauses 3(4), 10(3) and 17(3) and where the Engineer has reduced the sums claimed by the Contractor on behalf of the Subcontractor.

Within the UK withholding money, under the HGRA, is termed a withholding notice and through Case Law must be submitted or resubmitted (if the case has been made earlier than the application for payment) to the Subcontractor **after the receipt** of the particular application and **before the due date for payment** or it can be overturned on procedural grounds alone.

There is provision in the express terms entitling the Subcontractor to interest if the Employer defaults on an Engineer's certificate or if the Contractor breaches the payment provisions in this clause.

These rights are contained within Subclause (3)(e) and (f) (and it should be noted that the right to interest is not automatic, as under the Main Contract) and starts seven days after the date of notification.

Subclause (5) brings the release of retention into line with the provisions of the Main Contract. Although the onus is on the Contractor to pay without a specific request, interest is only due as is outlined above seven days after notification, so effectively the Subcontractor must still chase retentions.

Subclause (6) entitles the Subcontractor to full payment at the latest within the three months of having finally performed his obligations under Clause 13. Interestingly the Subcontractor jeopardises all this if he refuses to do properly ordered maintenance works.

Subclause (7) makes it essential that all submissions for additional monies are before issue of the Maintenance Certificate.

Subclauses (8) and (9) give force to the provisions of the Housing Grants Act with the necessity for a detailed statement of the sums to be paid and withheld and how they are calculated before payment.

Clause 16: Determination of the Main Contract

If the Main Contract is determined the Contractor may, after written notice, determine the subcontract. In such an event the Subcontractor shall remove his men, equipment and materials not vested in the Contractor. The Subcontractor clearly needs the Contractor's permission for Subclause (2) to be operable.

Subclause (2) provides for the Subcontractor to be paid for all materials, all work executed on or off site, and the reasonable costs of withdrawing from site. This is subject to the Subcontractor not being in breach of the subcontract prior to the determination.

Subclause (3) provides the further caveat on the payment provisions of Subclause (2) in that the Subcontractor is not the cause of the determination.

None of this helps the Subcontractor if the Contractor is made bankrupt or ceases trading.

The Employer is still able to impound all the Subcontractor's plant, materials and equipment through the provisions of the Main Contract.

As there is no contract in existence between the Employer and the Subcontractor the Employer is not required to pay for work executed or materials brought to site. This is why one sees a mad scramble to remove equipment and materials the moment the Main Contractor is likely to cease trading. Contractually incorrect, but self-preservation rules.

Clause 17: Subcontractor's Default

This clause gives teeth to all the powers given to the Contractor through the previous clauses. Examining the conditions upon which the sanctions can be applied we find again that, other than the act of bankruptcy (Subclause (1)(d)), the instructions to the Subcontractor must have been in writing. The grounds are if the Subcontractor in Subclause (1):

(a) suspends performance of his obligations under the Subcontract;
(b) fails to proceed with the Subcontract Works with due diligence after being required to do so in writing;
(c) fails to execute the Subcontract Works or to perform his other obligations in accordance with his subcontract after being required to do so in writing;
(d) refuses or neglects to remove defective work or make good defective work after being required to do so in writing.

If the sanctions that are considered below are to be applied it is legally safer to specifically warn the Subcontractor, obviously in writing, that unless he takes the required action by a certain time, which should be reasonable with regards to the task or requirement, then one of the courses of action outlined below will be taken, in accordance with the provisions of Clause 17. The Subcontractor is not entitled to a separate warning, provided the original instruction with which the Subcontractor has not complied was in writing and stated a time for compliance.

The Contractor can be faced with the problem where both parties are fully aware that the Subcontractor is not going to comply with an instruction. The Contractor must be in a position to operate Clause 17(3) (see later) and protect himself against claims for being in breach of the subcontract. However, he may not be able to afford the delay of waiting the reasonable period for the Subcontractor to comply. A solution is for the Contractor to write to the Subcontractor and, if necessary, deliver by hand an instruction stating that he requires an unequivocal written undertaking within 24 hours that the Subcontractor will comply. When a negative reply or no response is received within the time limit the Contractor is safe to proceed with the actions outlined below. It can also be seen why I recommend under the Clause 6 analysis that reference to Clause 17 be made in the standard letter to Subcontractors when ordering commencement.

Subclause (1) goes on to allow the Contractor, on giving written notice, to forthwith determine the Subcontractor's employment. This means a second letter stating what action is actually being taken. Legally it is necessary because the previous letter was a statement of intent and this second letter is the instrument of determination of all or part of the subcontract. To determine a significant subcontract the grounds must be very strong. I would always envisage first using the provisions of Subclause (3). If the Contractor still elects full determination he may confiscate all "plant, materials and things" brought onto the site by the Subcontractor, use them for completion of the works and/or sell them to off set against monies otherwise owed.

There are some practical and legal restraints on these far reaching powers. Firstly, sale of the Subcontractor's equipment etc. can only be effected if monies are owed to the Contractor, which may well be disputed. Secondly comes the question of ownership of plant and materials. The Main Contract deems that property in plant and materials pass to the Employer when on site. Clause 11 of the subcontract passes this requirement on to the Subcontractor. The subcontract organisation has normally set up a parallel plant company to the company which is in contract with the Contractor, both being owned by a central holding company. The plant is then deemed to be hired. Materials from most major suppliers and builders merchants are sold to the Subcontractor with the caveat that the supplier has a "deed of title" until he has been paid for the materials. If the Contractor is aware of such an arrangement he need not have paid the Subcontractor for materials on site until he had proof of ownership. Sale or incorporation of materials in the works without payment to the original supplier can be legally difficult.

Subclause (2) gives legal form to the action under Subclause (1). Subclause (3) introduces the measure in the CECA Form of Subcontract, the removal of part of the subcontract works from the subcontract and its execution at the Subcontractor's expense. The vital identical notices described earlier are necessary from the Contractor to Subcontractor. This is an invaluable instrument in the Contractor's armoury to ensure completion to time and quality and argue later with the Subcontractor without significant effect on the Main Contract. Subcontractors should note that the Contractor's costs for executing the works which they do not elect to do are likely to be considerably greater than their own. In refusing to execute work they gamble heavily. Their better course of action is to write, giving notice of the reasons that they consider the instruction invalid, stating that they are complying without prejudice to their right to claim additional monies and that records are being maintained for evaluation.

Clause 18: Disputes

The clause recognises the three procedures that are within the ICE Conditions of Contract, being conciliation, statutory adjudication (as provided for by the Housing Grants Act) and arbitration.

Subclause (2)(b) provides that any matter of dissatisfaction that, in the opinion of the Contractor, touches the Main Contract must be referred to the Engineer before it becomes a dispute. The Engineer must have the stated time to give his decision and a dispute only comes about when this has expired.

Subclause (3) brings in the possibility of conciliation provided a Notice to Refer to Arbitration has not been issued.

Subclauses (4) to (6) bring in the right to Statutory Adjudication using the Institution of Civil Engineers' Procedure. This is already the main dispute resolution procedure between the Contractor and Subcontractor. Although Adjudication is only a quick fix with the dispute to be decided by Arbitration very few disputes between Contractors and Subcontractors go beyond Adjudication. As far as Subcontractors are concerned, the introduction of Adjudication has been a great success shifting the contractual balance in favour of the Subcontractor. Before this, many a Subcontractor was bankrupt before the expensive Arbitration procedure had run its course; that very expense was also a deterrent.

The procedure mirrors that described under Clause 66 of the ICE Conditions and is not repeated here.

Subclauses 8 to 10 introduce arbitration which, again, follows the provisions of Clause 66 of the ICE Conditions and is not repeated here.

Clause 19: VAT

The law lays the duty upon the person or company charging for goods or services to collect VAT in the proper amounts. They are liable to pay the tax after issuing the invoice whether or not they have received the monies, except in the case of small companies under the 1987 finance act changes.

Currently all new building is zero rated, but some projects such as roads, may have realignment and widening section subject to VAT and a new section zero rated. The Contractor, as stated in the section on the Main Contract, should seek guidance from the local Inland Revenue Office if in any doubt. Subcontractors should seek their own decisions from the Inland Revenue unless the Contractor produces a photostat of the judgement obtained by him.

Clause 20: Law of Subcontract

This provides for the Law of the Country applying to the Main Contract to apply to the Subcontract. This is not just overseas, there are significant variations between English and Scottish law, particularly on the subject of vesting of materials off site.

THE NEW ENGINEERING CONTRACT SUBCONTRACT

Introduction

The New Engineering Contract Subcontract is virtually the same as the New Engineering Contract Main Contract and provides for the same combination of options with the names of the parties changed. For the basic provisions the reader can therefore refer to the analysis of the New Engineering Contract in the Contract Appendix. There are several reasons why Main Contractors are reluctant to use the New Engineering Contract Subcontract form which may be summarised as follows:

- Difficulty in providing accurate Works Information; the problem being in accurately splitting the various Main Contract provisions without causing a whole class of Compensation Events which have no counterpart in the Main Contract.
- Additional administrative burden; this can be considerable if the Contractor's various subcontractors are all calling him to early warning meetings and he has to answer all communications in a set time. The use and development of a whole series of different programmes to assess Compensation Events for different Subcontractors, in addition to those in use in the Main Contract, can make administration potentially impossible.
- There are fewer direct remedies where the Subcontractor fails to execute the work or any element of it in accordance with the programme.
- The caveats on the Contractor's services, attendances and remedies are eliminated or reduced.

Clause 26.3 of the New Engineering Contract requires the Contractor to notify the Project Manager of each set of Subcontract conditions to ensure that they permit the Contractor to meet his obligations under the Main Contract. Use of a different form, such as the CECA which is more onerous, is unlikely to result in rejection.

KEY DIFFERENCES FROM THE NEW ENGINEERING CONTRACT, MAIN CONTRACT

Completion

The Contractor has two weeks to certify completion on notification against the one week allowed under the Main Contract.

Take Over

In Clause 35.3, the Employer can use the Subcontract works.

Payment

- The Contractor has two weeks to certify, against one week in the Main Contract.

- The Contractor has four weeks from certification to pay, against three weeks in the Main Contract.

Compensation Event Notification

- The Subcontractor has one week to give notice, against two weeks in Main Contract.
- The Subcontractor has one week to submit a quotation, against three weeks in Main Contract.

Adjudication

There are provisions for Subcontract disputes to be joined in a combined hearing on Subcontract matters only (Clause 91.2) or joined to disputes on the Main Contract (Clause 91.3).

Also, the Subcontractor has only three weeks against four in the Main Contract to give notice on the action or inaction of the Contractor.

Termination

Clause 94.1 gives the Contractor the power to terminate at will but it would be unwise to do this without the type of warnings required under the CECA Form of Subcontract.

Labour Appendix

CONTENTS

TYPICAL INCENTIVE SCHEME PREAMBLE

1. OBJECTIVES

(a) To provide operatives with the opportunity to increase their earnings by additional payment above the basic weekly wage for work done in relation to predetermined performance targets.
(b) To provide management with means for increasing productivity and controlling production costs whilst maintaining high standards of workmanship and safe working conditions.
(c) To create a good working relationship between the Employer, Employee and the Trade Union.

050275

2. GENERAL PRINCIPLES

(a) Potential Earnings

No limit will be placed on bonus earnings but it is to be understood that to achieve such earnings there must be full acceptance that:

    (i)   Materials shall not be wasted.
    (ii) Quality standards are to be maintained.
    (iii) Safety Regulations are to be observed at all times.
    (iv) Resources will be deployed by management to produce the optimum efficiency of construction methods.

(b) Basis of the Scheme

    (i)   The scheme will be on a man hour basis. The savings between the actual hours worked and the total target hours will be paid at the bonus calculator rate of £X for Carpenters, Steelfixers and Scaffolders and £Y for General Operatives.

        Measure x Targets = Total Trade Target hours
        Total Trade Target hours - Total Trade Allocated hours = Hours saved
        Hours saved x Bonus Calculator rate = Total Bonus earned

        Where non-targetable time is allowed, this will be added to the total at the rate of £Z and will be deducted from the total allocated hours prior to the above calculation.

(c) Targets

    (i)   The targets within this scheme are inclusive of all special circumstances of access, storage, cranage limitations, etc. They are not applicable or comparable to any other schemes within the Company.
    (ii) Targets are fully comprehensive and have been set to include all items normally carried out on a construction site and those peculiar to this project, whether described or not.

        No further trades' targets will be issued; all allocated hours will be set against the items measured.
    (iii) However, if a class of work which, in the opinion of the management, is totally unrelated to issued targets occurs, this will be studied over a reasonable period and new targets issued.

        In the meantime work will continue and management will agree a temporary hourly rate which will be adjusted on issue of the new targets.
    (iv) A learning curve will be applied for the following reasons only:
        (a) For new gangs to the site.
        (b) For operations which are unusual within the industry. The amount of the learning curve to be a 20% easement in the first week and a 10% easement in the second week with no further additions thereafter.

3. MEASUREMENT OF WORK

(a) All measurements will be related to the drawings and schedules. Where site dimensions are taken for interim payment purposes, these may be adjusted to tally with the total on the drawings.

(b) Any measurement above that on the drawings will be verified by site engineering staff.

(c) Records of measurements taken shall be kept in such a manner as to enable quick and accurate reference to them to be made in respect of a query.

(d) An arithmetical check of the bonus calculations may be made by the appropriate steward at an agreed time each week. There will be no involvement of chargehands or gangers in the vetting and calculation of bonus payments.

4. QUALIFYING CLAUSES

(a) Bonus calculations will be made weekly and will include, as far as is practicable, all work done in the week.

(b) Bonus will be paid weekly on the second pay day following the completion of the work.

(c) If work has to be repeated or additional work done due to bad workmanship, then no target will be allowed for that work and all hours will be set against the accrued hours for the appropriate gang or trade in the case of a pool bonus.

(d) Overtime premium will not be set against a target.

(e) Hire charges of plant and equipment will not be set against bonus.

(f) Where welding is executed to assist in location of items for which the trades have targets, the welders' allocated hours will be set against the accrued hours for the appropriate section of the trade.

(g) Where the hours worked by plant operatives are included in bonus calculations, details will be included in the target preambles.

(h) Where an operative from Gang A is required to work in Gang B for a time he will be entered on the sheets of both gangs for the relevant periods and his bonus will be calculated accordingly.

(i) Foreman's time will not be charged against bonus.

(j) Non-productive time is defined as time in which the Company was not able to provide work for the men and the men were actually idle. This will only be allowed when the appropriate supervisor was informed at the time and subsequently approved the hours.

(k) Allocation Sheets are to be submitted to the Foreman by morning tea break of the following day. Failure to do so will result in management fixing the bonus rate.

(l) Special Cases

(i)   The following will not qualify for bonus:
- Chainmen
- Office and Canteen Attendants
- Latrine Attendants
- Bus Drivers (when driving bus).

(ii)   Hired lorry drivers will receive £Z per hour worked.

(iii)   Directly employed crane drivers will receive site average per hour worked.

(iv)   Externally hired crane drivers will receive 65% of site average per hour worked.

(v)   Van drivers will receive £Z per hour worked.

(vi)   Stores labourers will receive £Z per hour worked.

(vii)   The sawyer will receiver joiners' average per hour worked.

(viii) Pumpmen and fitters' labourers will receive £Z per hour worked.

(ix)   Fitters and welders will receive site average per hour worked.

(x)   Transport Organisation
- Hiab drivers (with HGV) 90% site average
- Hiab drivers 85% site average
- Tractor drivers 80% site average
- Labourers 75% site average.

(xi)   Directly employed excavator drivers will receive 95% of site average per hour worked.

(xii)   Externally hired excavator drivers will receive 55% of site average per hour worked.

(xiii) Crane Banksmen will receive 55% of site average per hour worked. (Banking of other plant items is included within the targets.)

(xiv)   Hired welders will receive £Z per hour worked.

5.   TERMINATION OF THE SCHEME

This scheme can be terminated by either side giving two weeks' notice of their intention to do so.

6.   THE AGREEMENT

This agreement does not in any way constitute a legally enforceable contract between the parties concerned. (This latter statement is imperative because no labour bonus scheme would stand the scrutiny of lawyers and unionised labour in the UK has ready access to law. Any incentive scheme is only suitable for particular phases of the project).

TYPICAL PROBLEMS IN OPERATING A BONUS SCHEME ON LARGE SITES

1.   Off booking of hours

(a) On to loss making targets

When certain targets are under attack as being inadequate for a fair recompense it may be suggested that the losses should not be carried forward. This should be resisted.

EXAMPLE

	Col A Quantity	Col B Rate	Col C Allowance	Col D Hours taken	Col E Saving
Formwork	$20/m^2$	$2mh/m^2$	40	25	15
Pump House Fix bolts	5 No	1 mh/No	5	6	-

If this is conceded then on this example the bonus for distribution would be 15 x X = £15X.

Now in a week's time following such a concession for the same work the gang might manage to book 20 hours to Formwork and 11 hours to bolts. The bonus would now be (40-20) x X = £20X. All of this can be resisted by insisting that the hours saved remain Column C total - Column D total, i.e. 14 hours at £X, i.e. a total of £14X which is the true reward available within the scheme.

The first of these examples gives a 7.14% increase in bonus and the second, when the men have started to move hours across onto the loss, makes an increase of 42.85%.

(b) On to non-productive time

The scheme should be set up to avoid recognising non-productive time, as described in the Typical Incentive Scheme preamble earlier in this Appendix, which covers all these items and was used successfully on a major site. However, if it recognises items such as waiting for Engineers to give levels, delivery of she bolts, walking to joiner's shop to collect wedges not delivered and the like, the following is liable to happen, particularly when initially the NPT hourly rate does not appear to be high:

	Col A Quantity	Col B Rate	Col C Allowance	Col D Hours taken	Col E Saving
True Situation					
Formwork	$20/m^2$	$2mh/m^2$	40	25	15
Pump House Fix bolts	5 No	1mh/No	5	5	0

NPT		1		-
				15
Allocation				
Formwork	20/m^2	2mh/m^2 40	23	17
Pump House				
Fix bolts	5	1mh/m^2 5	3	2
NPT			5	-
				19

With NPT being allowed at £Z per hour the following bonus would accrue if the true situation was registered.

Bonus for distribution = 15 x Y + 1 x Z = £15Y + Z
However, as allocated, the calculation becomes: (19 x Y) + (5 x Z) = £19Y + 5Z

Hence by a minor move of hours the bonus can be increased by up to 30% depending on the rates applicable. The standard way to stop all forms of off booking is for allocation sheets to be checked daily by section foreman and signed. This leads to endless arguments as the rewards to the operatives for squeezing a few more hours are considerable and the operatives will operate brinksmanship over a few hours. If conceded the front line managers consider confidence in them is reduced and the vigour of their checking diminishes and the reward to productivity ratio moves adversely. The fault is not in the front line management, but in the system which should be designed not to be affected by allocation as suggested later.

2. Pressure for Additional Targets

This can, of course, be legitimate if the Project Manager is issuing targets as the particular type of work crops up. It is often the case that a progressive issue of targets on a small site will be quite easy to control without any abuse occurring. However, on large sites, or where the labour force is known to be militant or organised, progressive release is dangerous.

The danger lies in the fact that, from the working of the particular target and other targets, it will be claimed that there is an area of work that the operatives are "not being paid for". If the scheme is growing all the time to meet the type of work encountered then the arguments will seem reasonable.

However, it becomes increasingly difficult to ensure that the sum of fragmented targets does not exceed the tender allowances. Obviously, if additional targets for items which should have been inclusive are issued then the effect on the productivity earnings ratio is greater than that for off booking.

The type of operation for which a target is issued is also important. The best is the all inclusive target related to the items which are measured with the client against which the operatives may, if necessary, draw a percentage weekly.

This means that the Bonus Surveyor can do all his measuring from the drawings, only visiting the site to check the mode of construction where this affects a rate.

3. Pro Rata

It will be surprising if at some stage the Project Manager does not have this request brought to him. It will be couched in reasonable terms stating that certain items in the preparation phase are too variable, or only particular items lend themselves to incentive schemes, or certain targets only are reasonable and the rest rejected. The request then is that the rate of bonus made on the targetable element should be applied pro rata to the other hours expended during the week.

Under no circumstances must this ever be conceded as the labour force can then milk the contract at will. An example is set out below for a welding operation:

Allocation Sheet Summary

Monday   4 hours splice piles, 2 hours walking time, 2 hours inspection
Tuesday   4 hours waiting for crane to turn piles, 2 hours repair to gate, 3 hours weld skip for concrete gang, 1 hour return suspect welding rods to stores.
Wednesday 2 hours weld pile bench, 3 hours splice piles, 2 hours await crane for turning piles, 1 hour awaiting Engineer.
Thursday   3 hours burn profile plates for fitters, 2 hours await transport gang to remove completed work, 1 hour welding lugs onto dumper, 2 hours splice piles.
Friday   3 hours working for fitters, 2 hours splice piles, 1 hour inspection by Clerk of Works, 2 hours wait crane to turn piles.

Hours spent on splicing piles =	11
Remaining hours	<u>31</u>
Total	42

3 No piles spliced at 15hrs/no	= 45 hrs allowance
Hence hours saved	= 45 - 11 = 34 hrs
Hence money earned in 11 hrs	= 34 x £Y = £34Y
Hence hourly rate	= $\frac{34Y}{11}$ = 3.09Y
Hence total bonus	= 3.09Y (11 + 31) = £129.78Y

This should be considered against what it is worth to the company, i.e.

$$45 \text{ hours} - 42 = 3 \text{ hrs}$$

Total available for bonus   = £3Y

Although this is an extreme example and off booking is also rife, it can result in this situation. Once the rate has been established by a short burst of work the object is to do as little as possible through legitimate excuses so that the golden eggs are laid for as long as possible and for the maximum number of operatives. No effort will be made to overcome any difficulty or report it to management as it is not in their interests.

If a Project Manager inherits a system where the principle of pro rata exists there are limited choices for corrective action:

(a) Repudiate the system unilaterally and sit out the consequent industrial action. If this course is chosen it should be recognised that the potential savings in bonus payments must be balanced by the costs of being late (i.e. increased fixed costs, acceleration costs and possible liquidated damages).
(b) If the system only applies to one trade then the site organisation can be adjusted to ensure that the operatives in that trade have higher levels of supervision and the resources of production so that they are always employed on targetable work.
(c) If the system has been conceded to the entire site and the option in (i) above is not feasible then the only remedy is higher staff supervision levels to enforce accurate hourly allocation of all hours, even though this in itself is liable to be a source of dispute.

4. Relationship between the Trade's and Labourer's Bonus

There are always difficulties if the bonus earnings of the tradesmen, and particularly the joiners, fall below the labourers' average. It is normally recognised that specialist gangs, like piling and concrete gangs, can earn similar amounts.

There is a very frosty reception when there is talk of fluctuations in the system and the possibility that perhaps the labourers have worked harder! The suggestion is not that the labourers should get less, but that if they have an earning potential at that level then a tradesman's must automatically be higher. This leads into the question of incentive schemes for labourers, as they cannot be considered in isolation because of the effects described above. The following are the possibilities:

(a) Produce a scheme with targets for every conceivable situation and operation. The disadvantages are the amount of work necessary pre-contract to prepare such a manual and the arguments that will occur in selecting the appropriate target.
(b) Have all tasks for the labour gangs extracted from the weekly programme and then get these evaluated by an experienced Bonus Surveyor. The advantage is that, presuming the evaluation is accurately carried out relating it to tender production rates, the labour costs are accurately controlled. However, the disadvantages are more numerous. Firstly, if you deviate from the weekly

programme or daily worksheet the ganger is off like a flash to have the new task assessed rather than doing and supervising the work. Secondly, the opportunity is there for labourers' average bonus earnings to exceed that for tradesmen with attendant difficulties. Thirdly, that the number of people required to run the scheme goes up.

(c) Produce targets only for the major bulk items for the contract such as placing concrete, laying drains and sewers and possibly kerbs. Other hours to be paid on a site average of some kind. The advantage is that only the specialist gangs on mainstream production have the opportunity of high earnings, the administration is less although, if the specialist gangs do general tasks as well, off booking must be watched for. The cost control, however, is less precise and the Project Manager can only look to the resourced budget programme to ensure that the correct number of men is keeping up the required progress with earnings no greater than forecast.

THE DISCIPLINARY CODE

This is a typical disciplinary code.

Stage 1   Any question shall be raised by the operative verbally with the foreman, provided that, where the operative's immediate supervisor is below the foreman, the operative shall first try to resolve the question verbally with such immediate supervisor.

Stage 2   Failing resolution with the foreman, the steward shall, if he considers there are good grounds for so doing, raise the question with the Project Manager or his nominee. The steward shall have the duty of reporting to the full time local official if the question is not resolved, and the full time local official shall, if he considers there are good grounds for so doing, pursue the question with the Project Manager or his nominee.

Stage 3   Failing resolution under Stage 2, the full-time local union official shall report the matter up to the appropriate union full-time official and the Project Manager (or nominee) shall report to the appropriate representative of the employer. Such union official, if he considers there are good grounds for so doing, shall pursue the question with such representative of the employer.

Stage 4   Failing resolution under Stage 3, the union official shall forthwith put the question in writing to the employer and it is the duty of such official and/or the employer to submit the question to the Board as quickly as practicable for settlement.

Where the question arising affects a category of operatives, they should raise it with their steward, who shall proceed direct to Stage 2 onwards. Where the question arising affects a number of operatives categories, they should raise it with their stewards, or, if appointed, the senior steward, who shall proceed direct to Stage 3 onwards.

Where no steward has been appointed, an operative may operate all stages of the procedure.

If it appears desirable at any stage of the foregoing procedure, with a view to maintaining productivity and the industrial harmony:

(a) the Board may intervene in accordance with Article Three of the Constitution; or
(b) either party to the dispute may refer the question to the Board before the procedure is exhausted.

Until the above procedure has been exhausted, there shall be no stoppage of work either of a partial or general character including a go-slow, a strike, a lock-out, or any other kind of restriction in output or departure from normal working. When a dispute arises, no action will be taken by either party to aggravate the situation.

The decisions of the Board shall be accepted and implemented by all concerned.

The crucial point is it is recognised that, in order to maintain good morale, the employer has the right to discipline those who fail to make appropriate use of the disputes procedure for resolution of questions arising without recourse to strike or other industrial action.

Claims Appendix

CONTENTS

This Appendix contains a series of issues where a source of guidance might be of use to the reader. The types of claims described are not normally encountered. They are:

- Rerating examples;
- Restitutionary claims;
- Breach of Implied Terms;
- Claims under Tort;
- Ex Gratia claims; and
- Global claims.

RERATING EXAMPLES

Example 1: Reduction of Quantity Without Affecting Nett Rate

A typical example under this heading would be formwork for a retaining wall which was not as high as anticipated; the number of uses are assumed to remain the same.
Assume

1. Declared spread of preliminary items, overheads and profit of 35%.
2. Quantity originally Xsq m now 12% less.
3. Rate = £y/sq m.

Calculation:

The total value of spread monies was: X x 0.35 = £0.35XY
The total for the item at = £0.88X x 0.65Y The nett rate = £0.572XY
The total including the = 0.35XY + 0.57XY tender spread = £0.922XY
(Hence by preserving the spread monies the reduction in total payment is 7.8% against a 12% reduction in quantity)

The new rate for the item would be $\frac{0.922XY}{0.88X}$ = £1.048Y /m2

Example 2: Reduction of Quantity Which also Affects the Nett Rate

The best example of this type is "handle pitch and drive" of piles. Assume that the Engineer has ordered the piles at the quantity in the Bill of Quantities, so payment for the material is not in question, and that only the handle pitch and drive rate is in dispute.

Further assumptions

1. Spread of preliminary items, overheads and profit declared at 35%.
2. Quantity originally X metres now 12% less.
3. Rate originally £Y /m.

Calculation: Spread Monies. The total value of the spread monies is:

$$0.35Y \times X = 0.35XY$$

Nett Rate:- The nett rate is built up from the labour and plant costed against the time for each pile. The time for each pile typically would be made up of 40% handling, 30% driving to the area for set and 30% driving in tightening conditions near to the set (some evidence of figures of this sort would be required either from the tender build up or a timed cycle on site). The reduction in length would only decrease the centre time portion by 12% as the other operations remain unchanged.
Hence the total for the item at the nett rate is:

$$0.65\,(.40 + (0.88 \times .30) + .30)\,XY = £0.627XY$$

The total including the Tender spread $0.627XY + 0.35XY = £0.977XY$

(Hence by adjustment of the varied aspect only and by maintenance of the Tender spread a reduction in total payment of 2.3% against 12% by quantity is achieved).

The new rate would be $= \dfrac{0.997XY}{0.88X} = £1.133Y$ /m

Based on the standard item inclusion for piling, the handling and disposing of any excessive off cuts of pile resulting from the reduced driven length would be added into the new rate as well, making it greater than the original rate.
The principle of this type of rerate is to look for as many items within the rate that remain unaltered by the reduction in quantity. For instance, in excavations the on/off costs of machines, ramping into excavations, pumping requirements and any other unaltered temporary works.
Most operations have a definable mobilisation phase and a finishing phase which is likely to remain constant, the decreased quantity affecting only the middle production phase.

Example 3: Reduction in Number of Uses of Temporary Works

A common occurrence involves formwork with the cause being a variation to part of the works. The variation will be priced in itself. The Contractor should agree and record in writing that a new separate shutter was required for the varied area to maintain programme. This will avoid argument at a later date on the basis that what was done was at the Contractor's choice and not essential or required by the Engineer. The valuation of the varied item, including the new shutter, will proceed easily, probably based on recorded costs.

The variation also has a knock-on effect on the unchanged element which does not have the number of uses of formwork included in the Tender.

Assume a wall within a structure was to be built in three lifts with there being only three uses of formwork. A series of slots and corbels were then introduced into the top lift. The following calculation rerates the lower two lifts. Again, a 35% spread of preliminary items, overheads and profit is assumed for consistency.

Calculation: the tender rate for the formwork for the full area was made up by distribution of the costs of make over three uses. Assume that the tender rate was £F/sqm which in turn was 1.352, where Z is the next nett rate.

At tender Z = (X hrs for fix and strike) + 0.33 (Labour + materials for make)
After variation Z = (X hrs for fix and Strike) + 0.5 (Labour + materials for make)

To this would be added the spread monies as described previously.

Example 4: Increased Quantities

There is obviously no case for applying for an increased rate if there has been no change in the method of working, resource levels or temporary works or any other direct cost which is a specific result of the increase. For an increased rate there has to be an effect covered by Clause 56(2), or a change in the character of the work as envisaged in Clause 52(1).

A clear case is the increase in depth of a suspended slab. Assume a previous slab thickness of X mm was increased by 50%. The area of framework would remain the same, but the props might also increase by 50% with increased bracing.

Tender nett = (Labour fix and strike soffit + formwork materials + Labour fix + strike falsework + falsework materials)

New nett rate = (Labour fix and strike soffit + formwork materials + 1.5 (Labour fix and strike falsework + falsework materials) + additional on/off costs of extra falsework materials)

To these nett rates would be added the spread monies as described in earlier examples to obtain the new rate.

RESTITUTIONARY CLAIMS

Restitution is the legal response to unjust enrichment of one party at the expense of another. It requires a benefit to the defendant gained at the expense of the claimant. As a general rule, restitutionary obligations will only arise where there are no applicable contractual obligations. Where there is a valid contract which deals with the issue, there may be little room for restitution. A very full and clear analysis is contained in Emden's Construction Law which can be paraphrased as follows.

Quantum meruit is the right to be paid reasonable remuneration. A quantum meruit claim may be based either on contract or on restitution. In principle, the conceptual distinction between contract and restitution is clear: contract is based on agreement between the parties; restitution is imposed by law. In practice, the concepts are often blurred together in the cases, whether because of muddled analysis or because the distinction makes no practical difference in the particular circumstances. The principal example of a contractual quantum meruit claim is where the contract between the parties is silent as to remuneration. In the case of building contracts, this situation is now governed by implied terms supplied by statute.

In summary circumstances giving rise to a restitutionary quantum meruit claim include the following:

- where parties proceed under what purports to be a binding contract, but is not;
- where services are performed by one party at the other's request but no contract for those services is entered into. This may arise where the parties intended to enter into a contract but never actually did so, or where the parties have entered into a contract but the services are outside its scope;
- where the contract has been frustrated; and
- where, before completion, the Contractor accepts a repudiation by the Employer as terminating the contract. In this event the Contractor can choose to sue either for damages for the breach or for a quantum meruit in respect of the work performed. The quantum meruit claim may result in a higher award if contract rates were low.

How the reasonable sum is to be determined depends on all the circumstances.

BREACH OF IMPLIED TERMS

The implied terms regarding the Contractor's performance are ignored in this analysis. The first examined are the terms that will be implied regarding payment where this is not clear or the Contract has not been finalised. The most usual situation where these provisions come in to play are in subcontracts not yet finalised. The second section is based on the Employer's duty to Co-operate and the third on the Unfair Contract Terms Act. Claims under this section are seldom made in isolation from the Contract clauses. The implied terms are there to define the duties under the Contract when for whatever reason they are unclear.

1. Payments

Within the UK these are imported by HGCRA 1996 where specific terms are not in the contract or not agreed.

- Entitlement to interim payments where agreed duration is greater than 45 days
- Period 28 days unless otherwise specified
- Amount of interim payments:

    a.   value of work performed in accordance with contract;
    b.   materials on site (if provided for in contract);
    c.   any other amount specified in the contract;
    d.   aggregate sums already paid or due as periodic payment.

Entitlement under this provision does not exceed contract price:

- Payment due later of:

    a.   7 days after the 28 day period:
    b.   the making of claim by the payee.

- Final payment
  Contract price within 30 days of completion or on application, whichever is later:
  Completion is not fully defined and it may be possible to deny on the basis of outstanding defects, etc.

- Period for payment: 17 days

2. Implied Terms Regarding the Actions of the Employer

Co-operation has been defined as positive and negative. Essentially the Employer must do what is necessary under contract and must not hinder.
    Examples of Negative Co-operation are:

- Issuing instructions which he has not authority to do;
- Effect of other contractors on the site. In the context of a contract between Contractor and Subcontractor the implied obligation not to hinder the Subcontractor was subject to the proper exercise of the Main Contractor's express power to regulate the timing and continuity of the works;
- Contractor entitled to execute the whole of the works;

- Similarly, the obligation of a Main Contractor not to hinder a Subcontractor would ordinarily extend to permitting the Subcontractor to complete the whole of the subcontract works.

Examples of Positive Co-operation

The Employer must therefore do those things which are necessary on his part to enable the Contractor to do the work. Examples include giving access to the site, obtaining permits, appointing an architect, giving instructions, and making nominations, all of which would have been required of him in a standard set of Conditions of Contract.

The duty of co-operation owed by a Main Contractor to Subcontractors is not absolute but may be affected by the contractual position of the Main Contractor and the Employer. The main contractor may be at risk of being delayed by matters which are outside his control and for which he has no right to recover compensation from the Employer. In those circumstances a Subcontractor cannot assert an implied term that the main contractor will continuously make work available so as to enable the Subcontractor to maintain efficient progress.

3. The Unfair Contract Terms Act 1977

The most important control over exclusion clauses and similar devices is that contained in the Unfair Contract Terms Act 1977. By this statute, certain varieties of clause are deprived of all effect, while others are given effect only in so far as they are shown to be reasonable.

The title of UCTA 1977 is somewhat misleading. A contract term does not fall within the statutory controls merely because it is unfair. Broadly speaking, UCTA 1977 governs only those contract terms which:

- purport to exclude or restrict one party's liability for negligence: UCTA 1977, s 2;
- purport to exclude or restrict one party's liability for breach of contract: UCTA 1977, s 3(2)(a);
- purport to enable one party to offer a different contractual performance from that to which the other party reasonably thought he was entitled, or to offer no performance at all: UCTA 1977, s 3(2)(b);
- purport to oblige one party to indemnify another against that other's negligence or breach of contract: UCTA 1977, s 4;
- purport to relieve one party from liability for breach of the terms implied into contracts for the supply of goods by the Sale of Goods Act 1979, ss 12-15 and the Supply of Goods and Services Act 1982, ss 2-5, 7-10: UCTA 1977, ss 6, 7; and,
- purport to exclude or restrict one party's liability in respect of misrepresentation: UCTA 1977, s 8.

CLAIMS UNDER TORT

Introduction

The Law of Tort is an enormous and complicated subject. I make no attempt at a comprehensive or theoretical statement of the law. Rather, the aim of this is to provide a summary of the Law of Tort viewed from the perspective of parties involved in or affected by building operations. Tort is a last resort and as it is not under a contract is not subject to Adjudication under the HGRA. This is High Court territory and the costs of actions will be immense.

Liability in tort may arise in a number of different ways in connection with building operations:

- for injuries to persons employed in connection with the building operations;
- for damage to personal property belonging to persons so employed;
- for injuries to the public (persons not employed in connection with the building operations);
- for damage to the personal property of the public;
- for damage to adjoining/neighbouring land and buildings;
- for economic loss caused by the building operations.

The two main causes of action in tort are negligence and breach of statutory duty. Their examination is followed by Defences to Tort Actions.

1. Negligence

The essential elements in establishing liability for negligence are:

- the existence of a duty to take care;
- proof of breach of that duty;
- proof of damage resulting from that breach.

It must be noted that the elements just described, if proved, will result in a party, A, being personally liable for damage caused by his negligence. It is also possible for A to be liable vicariously for the negligence of B. Whether A will be so responsible for the negligence of B depends on the nature of his relationship with B.

It is possible that A is made vicariously responsible for the negligence of B only where B himself is also liable to the claimant. The reason for considering whether A is also liable is usually a practical one: i.e. A is insured or solvent whereas B is uninsured and impecunious.

On this basis, therefore, it is usual to claim against the employer of a negligent person; the employer is very frequently vicariously liable.

Duty of Care

The question whether a duty of care exists can be a very difficult one. The most famous attempt at a statement of principle was made by Lord Atkin in Donoghue v Stevenson and resulted in 'the neighbour principle'.

The doctrine of the threefold test has developed to assess whether:

- the damage of which the claimant complains was a reasonably foreseeable result of the act or omission of the defendant called into question;
- there was a relationship of 'proximity' or 'neighbourhood' between the claimant and the defendant; and
- it was fair, just and reasonable to impose a duty of care.

The Extent of Recovery or Liability

The question of how far liability should be imposed for pure economic loss caused otherwise than by negligent advice frequently arose, in the context of building cases, in the 1980s and early 1990s. Pure economic loss means purely monetary loss: e.g. loss of profit as a result of inability to use premises. It is to be contrasted with monetary loss arising from physical damage: in this case the loss is of course measured in money (the cost of repairing the damage) but the loss is characterised as physical damage.

The present law therefore supports a distinction between a case where the defect causes physical damage to other property (or to persons), where a duty of care does exist, and a case where the defect is repaired before it causes such damage (where no duty exists).

There are three potential exceptions to this rule. The first is where the defendant does or says something which can be shown to be an assumption of responsibility to the claimant, enabling the court to impose a duty of care. The second has become known as the 'complex structure theory'. Where this second exception applies, the court recognises that the defect in question arises in one discrete part of the building, and goes on to damage the remainder.

In this event, the court re-characterises the damage suffered by the claimant as physical damage to the remainder of the building arising from the defective part. Thus a duty of care to avoid causing such physical damage can be imposed. The third is (perhaps) where damage to persons or property is imminent so that expenditure by the claimant is necessary to cure a truly dangerous situation.

In some circumstances it is possible to make a claim for the loss of a chance. The Court of Appeal in Allied Maples v Simmons & Simmons identified three categories of case:

    (a) where the defendant is guilty of a positive act or statement which was negligent. There the question is, what would have happened if the defendant had not so acted or advised?

    (b) where the defendant is guilty of a negligent omission to act or advise: there the question is, what would the claimant have done had the defendant acted or advised properly?

    (c) where the question is what would a third party have done had the defendant acted or advised carefully?

2.   Breach of Statutory Duty

Four principal questions must always be considered in relation to an allegation of breach of statutory duty:

    (a) does the law permit an action for damages for that particular breach?

    (b) has there been a breach, applying the true construction of the statute to the instant facts?

    (c) did the breach cause the accident?

    (d) are there any applicable defences and if so, what effect do they have on the claim?

In the first instance there would be a criminal action. In other instances the procedures above would apply.

Defences to Tort Actions

There are three defences to actions in Tort which can reduce the liability once it has been established. The first is founded in the 1945 Law Reform (Contributory Negligence) Act, which provides, by s 1(1):

> "Where any person suffers damage as the result partly of his own fault and partly of the fault of any other person or persons, a claim in respect of that damage shall not be defeated by reason of the fault of the person suffering the damage, but the damages recoverable in respect thereof shall be reduced to such extent as the Court thinks just and equitable having regard to the claimant's share in the responsibility for the damage..."

The second defence is based on the argument that a person may act recklessly and with full knowledge of the risks (created by the actions of himself and others) that he is running, so that he accepts (or is deemed in law to accept) those risks voluntarily. In such circumstances he cannot then raise an action complaining that another person was at fault in creating the risk and causing him damage.

    The third defence prevents the claimant from recovering where he sustained his injury as a consequence of unlawful conduct. However, it is not all unlawful conduct which bars a claim. The basis on which conduct will be determined to be sufficient to bar the claim is not easy to detect in the authorities.

EX GRATIA CLAIMS

This has happened in the Public Sector. On the Kessock Bridge Design and Construct contract in 1980 payments were made for additional material and fabrication costs for application of the Merrison Design Rules to deep plate girders. Although the Rules were in existence before the tender submission it was generally understood that they only applied to box girders.

GLOBAL CLAIMS

A claim based on all the various causes of the disruption and requesting "global" loss/expense, would not be struck out purely because the Contractor had not particularised the effect of each individual cause of the disruption. The global approach would be being applied to quantum and not the liability. This argument is strengthened where only the causes of disruptions that are the responsibility of the Employer are recorded and used in the quantification of the effects of disruption.

For anyone contemplating a Global Claim Lord MacFayden's statement in John Doyle Construction Limited v Laing Management Limited (2002) is of great importance:

> "A global claim, as such, must therefore fail if any material contribution to the causation of the global loss is made by a factor or factors for which the defender bears no legal liability. The point has on occasions been expressed in terms of a requirement that the pursuer should not himself have been responsible for any factor contributing materially to the global loss, but it is in my view clearly more accurate to say that there must be no material causative factor for which the defender is not liable."

Case Law appears to have developed the following seven rules/principles that apply to Global Claims if this hurdle can be overcome:

1. Global Claims may be used both in relation to loss/expense and time claims. A total costs claim is not automatically barred but is to be viewed with caution.
2. The Contractor must first establish that events have actually occurred which entitle him to the loss/expense or extension of time sought, i.e. the factual basis of each claim must be identified separately.
3. The Contractor must then show that he has complied with the contractual machinery for making claims in respect of each such event.

4. Where elements of the claim can practicably be isolated the Contractor must present claims in respect of these separately. A cause that could have been considered in isolation can be disregarded in assessing the global loss.

5. Even if the consequences of the individual heads of claim cannot be disentangled from each other, nonetheless the Contractor must still establish that each head of claim (or cause of disruption) qualifies for loss/expense and that each head did in fact cause (delay or) disruption.

6. The global approach can only be applied to quantum not liability.

7. The global approach is only justified in cases where it is difficult or impossible to make an accurate apportionment between interactive causes. The approach is one of last resort and cannot be used to lump all delay/disruption events together to justify a total cost and/or time overrun.

I would expect to present a Global Claim in exactly the same manner as described in Chapter 9, except the quantification that would be a Lump Sum. I would deduct from that sum a valuation of any known Contractor's risk element to ensure that it did not fall foul of Lord McFayden's ruling.

Contract Appendix

INTRODUCTION

It is no part of the purpose of this book to recommend particular forms of Main Contract because it is a guide for the Site Manager who has to operate with what he has got. The balance of risk is very different between the contracts.

Anyone using this appendix must first realise that **nearly every client alters the standard conditions** in the light of their experience and normally to alter the balance of risk within the published documents. Thus, **before consulting these pages look at the alterations**. It is a rule in the way contracts are interpreted by the courts that specific provisions take precedence over standard or general provisions. (It is still worth remembering that if the alteration produces a conflict within the document it will be read against the Employer.)

The Highways Agency who use the Highways Contract do not actually publish a standard edition, each project has a slightly different version, often eliminating the clauses regarding unforeseen ground conditions.

FIDIC has two parts (it is only Part I that is included within this analysis), with Part II allowing the Employer to tailor the contract to his needs.

The ICE Conditions are modified by Clauses 72 upwards, which often include, as well as additional provisions, alterations to Clauses 1 to 71. All of the above contracts can be altered and adjusted to different contract procedures, however the New Engineering Contract was designed from the outset to be adaptable to all the main procedures.

The New Engineering Contract consists of core clauses with a series of options. The main options defined by payment are:

- Option A - priced contract with Activity Schedule
- Option B - priced contract with Bill of Quantities
- Option C - target contract with Activity Schedule
- Option D - target contract with Bill of Quantities
- Option E - cost reimbursable contract
- Option F - management contract.

The secondary options are in 15 secondary option clauses labelled G to Z. Included within them are some matters such as retention and liquidated damages for late completion which most traditional contracts treat as essential. The full list is:

- Option G - performance bond
- Option H - parent company guarantee
- Option J - advance payment
- Option K - multiple currencies (for use with Options A and B)
- Option L - sectional completion
- Option M - limitation of design liability
- Option N - fluctuations (for use only with Options A, B, C and D)
- Option P - retention (for use only with Options A, B, C, D and E)
- Option Q - bonus for early completion
- Option R - delay damages (liquidated)
- Option S - low performance damages
- Option T - changes in the law
- Option U - special conditions (CDM Regulations in the UK)
- Option V - trust fund
- Option Z - additional conditions.

Even in this situation the drafting committee recognised that there would be clients who wished to change the balance of the contract and provided for Z clauses.

In this analysis only the core clauses are included unless specifically stated otherwise.

The ICE, FIDIC and Highways Conditions all have the same root and the analysis follows this. The analysis of the NEC, which was entirely rewritten by its drafting committee from first principles, is therefore disjointed. To simplify the analysis the following are left out which are identified by the ICE Clause:

3	Assignment
9	Contract Agreement
10	Sureties
15	Contractor's superintendence
16	Removal of Contractor's Employees on Engineer's request
17	Setting out
18	Boreholes and exploratory excavation
19	Safety and security
28 (1)	Patent rights
(2)	Royalties
29	Interference with traffic and adjoining properties
30	Transport of resources
33	Clearance of site on completion
32	Fossils
34	Rates of wages/hours and conditions of labour
35	Returns of labour and plant
37	Access to site
43	Time for completion

HOW TO USE THE ANALYSIS AND CONTENTS

The following are the headline items and compatible contracts:

5th Edition: also covers ICE Ground Conditions, 1st Edition and Irish 3rd Edition
6th Edition: Irish 4th Edition
7th Edition: also covers ICE Ground Conditions, 2nd Edition and ICE Term Contract
ICE D&C
FIDIC, Part 1
NEC 2nd Edition, core clauses

The analysis takes the form of a series of issues that are examined under the following headings derived from the ICE Conditions with cross references:

THE COMPARISON OF THE CONTRACTS BY SUBJECT

WORKS AND SITE INFORMATION

This concept is applicable only to the NEC where the responsibility for administration and design of the project is split. The definition is quoted in full.

(5) Works Information is information which
- specifies and describes the works or
- states how the Contractor Provides the Works and is either
- in the documents which the Contract Data states it is or in an instruction given in accordance with this contract.

(6) Site Information is information which
- describes the Site and its surroundings and
- is in the documents which the Contract Data states it is.

DEFINITION OF COST

5th Edition and FIDIC 1(5)
Cost: includes overhead cost on or off site.

6th and 7th Editions
Cost means all expenditure properly incurred or to be incurred whether on or off the site, including overhead finance and other charges which can be properly allocated.

DC
After other charges specifically includes in brackets loss of interest.

Highways [Clause 1.1]
Basically as the 6th, but does not include the word finance after overhead. This should not make any difference to recovery.

    The essence of this is that finance costs are only going to be certain on the 6th, 7th and Highways if the Contractor is actually paying interest on a bank account. Only the DC conditions specifically include loss of interest.

New Engineering Contract
This contract follows a different path with a fee (that is part of the Tender) being added to actual cost, the components of which are defined in the Schedule of Cost Components. Any item not in the Schedule is deemed to be in the fee.

THE EMPLOYER'S VARIOUS REPRESENTATIVES

It is fundamental to this comparison of clauses across the Contract to understand that the Engineer under the ICE 5th, 6th and 7th is a quasi-arbiter acting impartially. The ICE Design and Construct and the Highways Contract have Employer's Representatives and Employer's Agent which are not the same thing. The NEC takes a different position as set out below.

5th Edition: Clause 2 Engineer and Engineer's Representative
Provides through Clause 2(1) that the Engineer's Representative could not order work involving delay or extra payment. This could be modified by delegation under Clause 12(3) by the Engineer to the Engineer's Representative of all powers except decisions or certificates under Clauses 12(3), 44, 48, 60(3), 61, 63 and 66. It is up to the Contractor under this form to seek notification of delegated powers and to ensure that he checks instructions to see that they are under a delegated power.

6th Edition and FIDIC: Clause 2
Contains definitions of the Engineer's powers and at Clause 2(1)(b) recognises that the Employer may specifically limit these in the Appendix to the Form of Tender

or Part II of FIDIC conditions. Similar provisions exist regarding delegation of powers and limits upon them and at Clause 2(6)(c) the Contractor can request that the Engineer's Representative specify under which of his delegated powers an instruction has been given. The duty of impartiality is stated at Clause 2(8) (2.6 in FIDIC) and is limited where the Employer has limited the Engineer's powers.

7th Edition: Clause 2

Follows the 6th except that Clause 2(6)(c) is widened to encompass any person giving instructions under delegated authority.

DC: Clause 2

Follows the 6th Edition but without provisions for the option of limiting the Employer's Representative's powers to order variations and without the duty of impartiality. Approvals of technical data or drawings do not alter responsibilities.

Highways: Clause 2

The split is between the Employer's Agent, who is deemed to have full authority and an Employer's Site Representative. The latter has no intrinsic powers all of which must be delegated in writing. A seven day period must elapse in oral instructions confirmed in writing by the Contractor before they are deemed to be instructions. Again, approvals do not alter contractual responsibilities.

New Engineering Contract: Clause 11

Clause 11.2, after identifying the parties to the contract as the Employer and the Contractor, the Project Manager, Supervisor, Adjudicator and Subcontractor are recognised as separate entities. This is a very important split which follows the Latham principles described in Chapter 1. Although the Contract is generally silent about the Designer this can be a separate organisation which will be identified in the Works Information (Note: Clause 27.1 says that approval of the Contractor's design will be from "Others".)

Clause 29 provides that the Contractor obeys instructions given by the Supervisor and Project Manager which are in accordance with the Contract.

APPEALS AGAINST DECISIONS OF DELEGATED PERSONS

5th Edition: Clause 2(4)

Permits any decision of the Engineer's Representative to be referred to the Engineer for his decision. There is no time specified for the Engineer to provide it.

6th Edition

Clause 2(5)(c), in addition, permits the appeal to the Engineer's Representative of any matter decided by a delegate person. The provisions of the 5th are now in Clause 2(7).

7th Edition

As 6th, but with appeal to Engineer if Engineer's Representative's decision is removed.

FIDIC, DC and Highways

These provisions do not exist.

New Engineering Contract

As with FIDIC, the provisions do not apply. However, adjudication is included which, outside the UK (where statutory adjudication is available), will permit decisions to be readily reviewed.

SUBCONTRACTING

5th Edition: Clause 4

The Contractor cannot subcontract the whole of the works and written permission is required to subcontract any element.

6th Edition: Clause 4

This Clause was revised to recognise that Subcontractors and the self-employed constitute the major proportion of the labour force in the UK. Subcontracting the whole of the Works requires the Employer's consent; subcontracting apart of the Works requires prior notification to the Engineer but no consent is necessary. Employment of labour-only subcontractors does not require prior notification. The Contractor is still fully responsible for all subcontracted work (Clause 4(4)) and the Engineer has the specific right "after due warning in writing" to require the Contractor to remove from the Works any Subcontractor who misconducts himself (Clause 4(5)). (This is additional to the powers given to the Engineer under Clause 16 to require the removal from the Works of any employee of a Subcontractor who misconducts himself.)

DC: Clause 4

Follows 6th but in subclause 2(a) the Contractor must obtain consent prior to making a change to the Contractor's designer named in the Appendix to the Form of Tender.

7th Edition: Clause 4

Follows 6th but in subclause 2(a) the Engineer may object "for good reasons" which must be stated in writing within 7 days. The Contract is silent on what happens if the Contractor ignores the Engineer's objection.

Highways: Clause 4

Clause 4(4) Named Subcontractors in the Appendix cannot be removed without written approval.

Clause 4(1) No Subcontractor may commence work until the Contractor has submitted a certificate with the following information and had it returned by the Employer's Agent as "not objected to". This must state:

- identity and scope of package
- confirm that Subcontractor selected because he can meet ISO 9002 (see Chapter 12) and his financial obligations
- the Subcontractor, if appointed, will operate a quality system to Clause 28
- the ability to meet obligations by list of previous contracts.

No time scale is given for the Employer's Agent to grant this certificate.

FIDIC: Clause 4

Follows the 5th but provides that no approval is necessary for any subcontractor named in the Contract.

Subclause 2 states that any residual obligation to the Contractor by the Subcontractor beyond the Defects Liability period shall be assigned to the Employer at his request.

New Engineering Contract

The provisions cannot be easily paraphrased and are therefore quoted in full.

26.1   If the Contractor subcontracts work, he is responsible for performing this contract as if he had not subcontracted. This contract applies as if a Subcontractor's employees and equipment were the Contractor's.

26.2   The Contractor submits the name of each proposed Subcontractor to the Project Manager for acceptance. A reason for not accepting the Subcontractor is that his appointment will not allow the Contractor to Provide the Works. The Contractor does not appoint a proposed Subcontractor until the Project Manager has accepted him.

26.3   The Contractor submits the proposed conditions of contract for each subcontract to the Project Manager for acceptance unless:

- the NEC Engineering and Construction Subcontract or the NEC Professional Services Contract is to be used; or
- the Project Manager has agreed that no submission is required.

The Contractor does not appoint a Subcontractor on the proposed subcontract conditions submitted until the Project Manager has accepted them. A reason for not accepting them is that,

- they will not allow the Contractor to Provide the Works; or

- they do not include a statement that the parties to the subcontract shall act in a spirit of mutual trust and co-operation.

The Project Manager has to reply within the period stated in the contract.

COMPATIBILITY OF CONTRACT DOCUMENTS

5th, 6th and 7th Editions: Clause 5: Contract Documents
The clause gives the Contractor the right to expect that the documents are consistent throughout. See also Clause 13(3) which provides for extra payment to the Contractor caused by disruption under Clauses 5 and 13. Practically, it is often hard to decide whether the difficulty requires further information or is an actual discrepancy. If in doubt Clause 5 is more advantageous as it requires the Engineer to respond immediately by the use of the word "thereupon" which is not present in Clause 7. There is a common law duty on the Contractor to mitigate cost and delay and he should not sit back. There is always the possibility that the explanation will not require instructions.

DC: Clause 5
Ambiguities between the Employer's Requirements and the Contractor's submission are to be decided by the Employer's Representative. However, the Employer's Requirements take precedence, which limits the scope for recovery.

The Employer's Representative's instructions are to be valued under Clauses 44, 53 and 60.

Highways: Clause 5
There are two additional caveats to those for the ICE DC contract that are important. The first is in 5.3 which excludes discrepancies that the Contractor "ought reasonably to have discovered in the tender period". The second is in 5.4 which states that the Contractor has only 28 days to dispute any ruling on technical interpretation which has to go to the Adjudicator. Despite there being substantial hurdles this is only the **third clause and the fifth cause leading** to an Employer's Change without one being instructed.

FIDIC: Clause 5
Because of its use in many areas of the world, Subclause 1 begins by stating that Part II will contain the language, country or state in which the law shall be interpreted.

The first part of Subclause 2 follows the ICE 5th to 7th but adds a document hierarchy, being:

1. Contract Agreement
2. Letter of Acceptance

3.   Tender
4.   Part II of the FIDIC Conditions (which are Contract specific)
5.   Part I of the FIDIC Conditions, which are analysed in this book
6.   Any other document forming part of the Contract.

It is unusual in engineering contracts for the Tender to take this position.

New Engineering Contract: Clause 17

Both the Project Manager (who is not necessarily the Designer) and the Contractor notify each other of inconsistencies or ambiguities in or between the documents which form the Contract and the Project Manager gives instructions to resolve them.

Clause 63.7 regarding assessment of Compensation Events provides that, if the Employer provided the data that is changed to remove ambiguity, the conditions most favourable to the Contractor is assumed. If the Contractor provided the data then it is reversed.

EMPLOYER'S DESIGN LIABILITY

5th Edition: Clause 6: Supply of Documents

The supply is put firmly as being the Engineer's responsibility. The contract drawings are issued after award of the contract and may differ in important matters from the tender drawings. Hence the Contractor should request the contract drawings immediately upon award, or a written statement that the tender drawings are to be the contract drawings. On major public works projects such as bridges, there are very few alterations; but on industrial projects the design is never stable and major changes can happen between tender and contract. In such cases the best solution is to submit the Clause 14 programme based on the tender drawings. If this is not done any effects of the revised information must be made clear somewhere in the submission.

6th and 7th Editions: Clause 6

As the 5th, except that four rather than two copies are to be supplied and provision is made for four copies of drawings of any Contractor-designed elements to be supplied to the Engineer.

DC

Clause 5(2) provides for the Employer to provide one copy of the Contract Documents.

Highways

No similar provisions.

FIDIC
As 5th Edition, but where Contractor supplies drawings there must be four copies. Clause 6 rather than Clause 5 contains the entitlement to cost for non performance (see later).

New Engineering Contract
All that is necessary is deemed to be within the Works Information at the beginning of the Contract unless the Contractor is to design all the Works or elements of the Works.

FURTHER DRAWINGS AND INSTRUCTIONS

5th Edition: Clause 7
It is mandatory that the Engineer issue any necessary further drawings and instructions by use of the word "shall", the qualification of "in his opinion" can be arbitrated upon.

Subclause (2) requires "adequate notice in writing" to be given of the requirements. Until Merton v Leach, 1985 it was generally accepted that arrows or markings on the approved Clause 14 programme were not adequate. The procedure outlined below in the rules section and under starting up a project should still be put into effect whereby updated notices are continually given on a monthly basis, preferably immediately before the monthly progress meeting.

Subclause (3) allows payment through Clauses 51, 52(4) and 60 with possible extensions of time through Clause 44 from the Engineer's failure or inability to provide further information. Even when the proper notices have been given it is often difficult to prove effect. On the one hand, the Contractor has a duty to mitigate the effect of delay; on the other, if he does this by winding down his operations the Engineer may remain silent throughout the process and produce the necessary further information at a time which matches the reduced work rate the Contractor has adopted. The Engineer could then claim that there was no effective delay. The action being taken by the Contractor must be clearly recorded in writing to protect the interests of the Contractor.

6th, 7th Editions and FIDIC: Clause 7
As 5th, but with similar provisions for the Contractor to supply details of any Contractor-designed elements.

DC
Similar to the 6th and 7th but in Clause 6(1)(a) and (b) after provision by the Employer's Representative of further information there is access to time and cost if he does not.

Highways
No similar provision. Information will have to be gathered via Clause 5.

New Engineering Contract
No separate provisions from those identified under "Compatibility of Contract Documents". See also Clause 20.1.

REQUIREMENTS ON THE CONTRACTOR REGARDING DESIGN POST CONTRACT

5th Edition
No provisions in standard clauses.

6th and 7th Editions
Examined in clauses above where similar provisions to those for the Engineer apply to design and also in Clause 8.

DC
Clause 6(2)(a) to (c) provide for the process of submission and approval of the Contractor's design. Subclause (d) gives access to recompense and time if approval is unreasonably delayed. Copyright is ceded to the Employer under Clause 7.

Highways
Clauses 6 and 7 contain a very prescriptive design process defining Designer, Checker, Safety Auditor and Quality Certification. There are no provisions for payment of additional cost in any delay in acceptance of the design. Clause 8.2 emphasises this and Clause 8.6 requires Design Certificates from the Contractor's Designer to be accepted before work can start.

New Engineering Contract
Clause 21 provides that the Contractor designs what is stated to be designed by him in the Works Information to the standard set out therein. The liability for design will be capped within the Contract data. Reasons for rejection are non compliance with the Works Information or the applicable law.

Approval of the Contractor's design is by "Others", Clause 27.1, which caters for a separate designer.

GENERAL OBLIGATIONS

5th Edition: Clause 8: General Obligations [Also 8 in Highways]
Clause 8(1): Contractor to provide everything necessary to complete the works. Provides in 8(2) that the Contractor is not responsible for the Permanent Works' design and any Temporary Works' design executed by the Engineer.

FIDIC, 6th and 7th Editions

As the 5th Edition but recognises that some Permanent Works elements may be designed by the Contractor and that his liability is to use all reasonable skill, care and diligence.

Highways

As 5th in Clause 8.1. Subsequent subclauses deal with the Contractor's design.

DC

As 6th and 7th but also specifically requires in 8(2)(b) that any element of the Works designed on behalf of the Employer must be checked and accepted as the Contractor's design.

New Engineering Contract

Provision to complete the works in Clause 20.1. See comment above regarding Contractor's design liability.

QUALITY ASSURANCE

5th, 6th, 7th Editions, FIDIC and DC:

The ICE 5th, 6th, 7th and FIDIC do not include requirements for QA plans. The Design Construct contract in Clause 8(3) leaves these matters as optional and to be specified within other contract documents. If required delayed approval will result in recovery.

Highways: Clause 28

Mandatory systems to comply with ISO 9002 and no provision for delay.

New Engineering Contract

Not specified in the Contract Document itself, but if required would be in the Works Information.

INSPECTION OF SITE AND SUFFICIENCY OF TENDER

In the UK the Contractor is permitted to rely upon information provided as part of the Employer's duties under CDM (see Checklist on page 89) which is especially important when contracts involve working in or upgrading existing facilities.

5th Edition: Clause 11: Inspection of Site

The inspection of the site and sufficiency of tender will usually have been settled before the site team is assembled.

Subclause (1) and the 1967 Misrepresentation Act entitle the Contractor to assume that soils and borehole information provided is accurate. Disclaimers as to the accuracy of, and responsibility for, such information are void.

It is assumed that the Contractor's inspection of the site prior to tender was reasonably thorough and conditions which should have been observed will not lead to the ability to claim additional monies.

Subclause (2) ensures that the tender is inclusive of whatever is necessary to execute the works in a proper manner in accordance with the contractual obligations and for the price quoted, warts and all!

6th Edition: Clause 11

The rewriting appears to have followed case law. The new Clause 11(1) means the Employer is deemed to have made available all existing relevant site investigation data. At 11(3) the Contractor is deemed to have based his tender upon it.

7th Edition and DC: Clause 11

As 6th with the additional specific provision that the Employer is to provide in 11(1)(b) details of cables and pipework across the site. (Also accords with CDM Regulations which actually goes further).

Highways (Clauses 11 and 12)

11.1.1, 11.1.2 and 11.2 leave the Contractor with the entire risk for satisfying himself as to the nature of the site. Some Employer's Agents are interpreting Clause 11.2 as precluding any claims under Clause 5 or any of the other exemptions. It is important to note that the clause commences with "Save as otherwise provided under the Contract."

Clause 11.3 eliminates adverse weather as a cause for extension of time. Clause 12 brings in sufficiency of tender which is in 11(2) of the ICE Conditions.

FIDIC: Clause 11

Clause 11 requires the Employer to provide all relevant investigative and survey data available; the Contractor is deemed to have based his tender upon it but the clause makes the Contractor responsible for its interpretation. The Contractor is also deemed to have made his own observations from examining the site and in particular the hydrological and climatic conditions which are not obtained by a single visit. Clause 12.1 introduces the Contractor's responsibility for sufficiency of tender in the same terms as the ICE Contracts have in Clause 11(2).

New Engineering Contract

The provisions come after Compensation Events and cannot be paraphrased and are:

60.2
In judging the conditions referred to in Clause 60.1(12), the Contractor is assumed to have taken into account:
- the Site Information;
- publicly available information including any referred to in the Site Information;

- information obtainable from a visual inspection of the site; and
- other information which an experienced contractor could reasonably be expected to have or to obtain.

60.3

If there is an inconsistency within the Site Information (including the information referred to in it), the Contractor is assumed to have taken into account the conditions referred to in Clause 60.1(12) more favourable to doing the work.

These have been designed both to be reasonable and to comply with Case Law.

UNFORSEEN CONDITIONS

5th Edition: Clause 12

This clause is the most used and abused in the Conditions of Contract and a minefield for the unwary. The clause is one of the few that **specifically includes Profit.**

Subclause (1) limits the action of the clause to physical conditions and artificial obstructions. In the majority of cases this means, for the former, variations from the soils information provided at time of tender and, in the latter, uncharted or incorrectly documented man-made objects. The latter being reasonably obvious, let us consider variations from the soils information. The subclause goes on to say that the condition "could not reasonably have been foreseen by an experienced Contractor."

Clause 11 discussed earlier allows the Contractor to place reliance upon the borehole information.

The Contractor is only concerned with conditions as found. He should not concern himself as to how foreseeable they were. If it causes a change in the intended tender methods, notification must be given. This can always be withdrawn later if the Engineer proves to the Contractor's senior managers and directors or that the condition was foreseeable or caused by the method of working. This protects the Company's position. With this notice, or as soon as possible thereafter, are required details of proposed measures and their anticipated effects. The Contractor should ask for approval of these under Subclause (2)(b) discussed below. Once this process is in train the procedures of Clause 52(4) must be followed by the Contractor (see discussion of this clause later).

Subclause (2): the courses of action set out in items (a) to (d) are entirely discretionary and the Engineer need do none of them. Normally it is only when the nature of the circumstances is so blatantly unforeseeable that he is likely to take these actions. In cases of doubt it is the Site Management's role to steer the Engineer and his Representative into approval of the measures taken. It is essential to get alongside the Engineer's Representative, discuss the problem, and get agreement for the measures taken. All engineering activity below ground is subject to varying opinions as to the best constructional methods in any given set of circumstances. It is conceivable that the Engineer's Representative could improve upon the Contractor's ideas! The main advantage remains that it cannot be said at a later date that "if the job had been done properly (i.e. differently) the effects would have been less." The Employer's interests are not damaged as there is consensus in engineering opinion.

Subclause (3) provides that the Contractor is due, subject to the Engineer's opinion, direct additional cost, and extensions of time with associated cost, for "unavoidable delay and disruption". The word "unavoidable" emphasises the need for engineering consensus on the methods of tackling the problem.

The other important point is that the subclause includes the following phrases "had such conditions or obstructions...not been encountered" and later on "suffered as a consequence of encountering the said conditions or obstructions." This is important in two fields often encountered.

Firstly, take the case of a weather-sensitive operation. (e.g. concreting) which is delayed for a week by unforeseen obstructions that, except for the delay, would have been executed in good weather, but is pushed into a week of freezing temperatures. As this is the result of the first delay it should be evaluated as a continuation.

The second area is the protection of the Contractor's float, either inherent in the programme or generated by successful earlier operations. Consider an operation which is shown on the programme as being of 15 weeks duration, and that it is 66% complete after seven weeks and takes a further 10 weeks to complete after a Clause 12 occurrence.

Many Engineers will say that the paid delay is $10 + 7 - 15 = 2$ weeks. However, had the circumstances not been encountered the Contractor, having only a third of the work remaining, could have reasonably been expected to complete in a further five weeks, rather than the 10 weeks taken. Hence, the cost which would "not have been encountered" is for five weeks and not two weeks.

Subclause (4): the interesting part of this subclause is that awards of time or cost given on an interim basis cannot be withdrawn. Often after receiving a notice under this clause the Engineer rejects the grounds for the claim. He may activate Clause 46 (which is discussed later) to maintain consistency of approach. The rules of action for this clause set out below assume a marginal case which has a good chance of recognition after a formal submission is made. In more serious cases the Contractor who disagrees with the Engineer's decision will not be able to challenge it in practice until he has presented a formal claim document which may be necessary to place him in a position to ask for an Engineer's decision under Clause 66 or to ask for Adjudication in the UK.

Summary Advice for 5th Edition

(a) The Contractor should write to the Engineer informing him of the circumstances, and claiming under Clause 12(1) giving details of the proposed measures to be taken and requesting approval of the measures under Clause 12(2)(b). If this is given the case proceeds smoothly.

(b) Assuming silence on the part of the Engineer the Contractor should get alongside the Engineer's Representative and discuss the occurrence and the best method of tackling it. At the minimum, the Contractor should be able to convince the Engineer's Representative to accept a letter which details the discussion and ends "we confirm that, although you maintain that the conditions were foreseeable, the methods proposed are the best solution in all the circumstances."

(c) Maintain records meticulously.

(d) Agree all subsequent alterations using the same phraseology as (b) above.

6ᵗʰ Edition: Clause 12

The previous Clause 12(1) has been split into three separate clauses. Clause 12(1) now calls for the earliest possible written notification to the Engineer of a Clause 12 situation. Clause 12(2) requires separate (or simultaneous) notification of the Contractor's intention to claim additional payment or extension of time. Clause 12(3) requires the Contractor, when giving notification under either of the previous two clauses (or as soon as possible thereafter), **to give details of the anticipated effects, the measures he has taken or is proposing to take and their estimated cost and the possible delay or interference with the execution of the Works**.

The other additional provision is that the Engineer may require the Contractor in Subclause (4)(a) to investigate the cost and timing of alternative measures.

Both these new clauses throw a considerable burden on the administration of the Contract. In effect, a full rerun of the Clause 14 programme network with all prolongation of activities will be required to ensure that all potential effects are costed and identified. If there is significant delay on the part of the Contractor in producing information for the Engineer to decide and instruct upon, this element will not be included in the award.

Despite the intention that the Engineer should act it is still possible that he will not and previous advice remains. To this should be added:

Additional Summary Advice for the 6ᵗʰ Edition
Ensure that a rerun of the Clause 14 programme network has been done and all prolongation of other activities included within the estimates.

7ᵗʰ Edition: Clause 12

Follows the 6ᵗʰ but in Subclause (3) the reference to "Subclause 1 and/or 2" considers a possibility where a Contractor may notify an obstruction or adverse physical condition but not intend to claim! In this unlikely event the same cost and time effects are still to be provided. Through modifications to Subclause (6) separate notice for time under 44(1) is not required and the Engineer must consider time under Clause 44(3) forthwith.

2nd Edition Ground Investigation Conditions: Clause 12 and 12A

In addition to Clause 12, which follows the 7th, there is a new clause 12A which deals with the provisions for unforeseen contamination discovered during a ground investigation.

This could refer to contamination that was in excess of that provided for in the contract as well as contamination where none was expected.

The radical new requirements are in clause 12A(1) which place the duty on the Contractor to notify the relevant authorities and to decide independently of the Engineer whether to suspend works for the safety of "persons and property or otherwise" to suspend all or part of the site operations.

Subclause 2 then provides for the Engineer to consult with the relevant authorities and the Contractor and then to either confirm the suspension or take action under Clause 12 (5).

Subclause 12A(3) provides that any suspension confirmed by the Engineer will be deemed to have been ordered under Clause 40(1) by the Engineer. However, it is silent on what happens if it is not confirmed. It must be hoped that the Engineer will still certify in this eventuality under Clause 12A(4) which states that "any delay or extra cost" shall be considered under Clause 12.

DC: Clause 12

The structure of the clause follows that of the 6th. Given that it is design and construct the inclusion of fully costed alternatives is left to the discretion of the Contractor (included in extended Subclause 3 rather than Subclause 4(a) of the 6th), and the Employer's Representative does not have the power (Subclause 4(c) of the 6th) to instruct how the conditions are to be dealt with.

Highways: Clause 13

In this Design and Construct version the relevant Clause is 13 and is usually excluded, thus the entire risk of unforeseen circumstances is passed to the Contractor. Where Clause 13 has been used it includes a caveat regarding the positioning of all Statutory Utilities apparatus.

FIDIC: Clause 12

Clause 12(1) contains the provisions of sufficiency of Tender which are in Clause 11 of the other ICE-based contracts. Subclause (2) is a throwback to earlier ICE conditions. There are no time limits on notices.

The Engineer is not expected to act; indeed, the determination of cost is to take into account any reasonable measures acceptable to the Engineer which the Contractor may take in the absence of specific instruction from the Engineer. The Engineer is to consult the Employer before awarding time and cost (which is a warning).

New Engineering Contract

The provisions are within Compensation Events and not in a separate clause. The provision is that if there is an inconsistency within the Site Information (including the information referred to in it), the Contractor is assumed to have taken into account the physical conditions more favourable to doing the work.

The requirements regarding notice are described in Notice and Claim Procedures and crucially comprise the Early Warning Procedures of Clause 16 which are described there.

METHOD RELATED PROVISIONS, LEGALITY AND IMPOSSIBILITY

Before commencing this review it is worth considering the concept of impossibility as it applies to construction contracts as it arises within all of the contracts reviewed below. It does not mean that the matter has to be absolutely impossible (indeed given that space travel is possible it would negate the intention if this was so). A matter is deemed to be impossible if it cannot be achieved by conventional techniques or the required items purchased on the open market.

A further important point is that Case Law has established that specification relaxations which remove the impossibility are also variations that should be valued.

5th Edition: Clause 13: Work to be to the Satisfaction of the Engineer, Mode, Manner and Speed of Construction

Subclause (1) gives the Engineer the power to instruct or direct on any matter connected with the Contract and significantly makes it clear that only he (or subject to Clause 2 the Engineer's Representative) has that power.

Under the Contract the Employer has no power to instruct the Contractor directly. He must give all instructions through the Engineer. The Engineer cannot pass on to the Contractor any such instructions which contravene the Conditions of Contract. The Engineer's duty is to arbitrate between the two parties to the Contract.

There is the extremely important caveat in the first line, excluding matters which are legally or physically impossible. Variations are required to remove the impossibility (see opening comments under this heading).

Subclause (2) brings in the matters which can be specifically varied by the Engineer or, more likely, will be affected by a variation being mode, manner and speed of construction.

Subclause (3), as well as catering for instructions given under Clause 5, provides effectively for all elements which were unforeseen and are not physical or artificial obstructions, providing that they are instructed or directed by the

Engineer. Hence it is necessary for the Contractor to refer all such matters to the Engineer for his instructions. The Engineer will not necessarily oblige with an instruction stated to be given under Clause 13(1). The Contractor then has the opportunity to say that the requirements of the Engineer are being executed on the basis that they are effectively instructions under Clause 13(1) to be valued through Clause 13(3). Eventually the Site Team or someone within the Contractor's organisation will have to prove that it was "beyond that reasonably to have been foreseen by an experienced Contractor", but the case is presented in the best possible manner.

Summary Advice for ICE Conditions

(a) Accept instructions only from the Engineer and, subject to Clause (2), the Engineer's Representative. If Clause (72) introduces any other authority confirm its instructions to the Engineer.

(b) Any matter which was unforeseen at the time of tender should be taken to the Engineer for his directions and then these should be confirmed back to him as "constituting instructions under Clause 13(1) which have disrupted the Contractor's arrangements and hence an extension of time with costs is claimed pursuant to Clauses 13(3), 44, 52(4) and 60."

6th Edition: Clause 13:
Follows the 5th Edition except that profit is added to cost.

7th Edition: Clause 13
Follows the 6th Edition, except that instructions under 13(1) can be given by any person with delegated authority.

FIDIC: Clause 13
Part I only contains the provisions in Subclause 1 of the 5th analysed above. This is important as the elements of variations are gathered in Clause 51 and include sequence and timing but do not include method or mode. The Engineer can still instruct but there is no variation for the latter item.

DC
Given that the Contractor designs there is no provision for the Employer's Representative to instruct on mode, manner and speed of construction. It is assumed that the Contractor will not design anything that is physically impossible to construct. Clause 13 is stated as "Not used". Legal impossibility is linked; the Employer's Requirements being legally impossible and are covered in Clause 26(3) leading to variation under Clause 26(3) (b) and 51, as discussed earlier.

Highways

Follows DC. The provisions are in Clause 32 and reimbursement is through Clause 32.1.2 and an Employer's Change through Clause 53. There is a distinction in Clause 32.1.3 to ensure that only statutory changes which affect the Contractor's proposals lead to change.

New Engineering Contract

Impossibility is covered by Clause 19 where the Contractor is to give notice as soon as he becomes aware of the Works Information requiring him to do something impossible or illegal. There are no provisions to instruct on methods, sequences and the like. However, Clause 34 permits the Project Manager to instruct the works to be stopped or restarted, which gives some control of the speed of construction if it should be necessary.

CONTRACT PROGRAMMES

5th Edition: Clause 14

Subclause (1) requires the provision of a programme "showing the order of procedure" and a "general description of the arrangements and methods of construction" within 21 days of acceptance of the tender.

It is highly advisable that the Contractor's tender intentions remain clearly demonstrable. Overall float should be marked as Contractor's float, filling the period up to section completions given in the form of tender.

Subclause (2) is titled in the margin notes as "Revision of Programme". It is essential that the Contractor does not issue programmes to the Engineer or the Engineer's Representative which are not either the Clause 14 programme or programmes ordered specifically under Clause 14(2) unless the Contractor's directors have accepted that the Contractor is at fault. Working to revisions that are not ordered leaves the Contractor without the protection provided by the contract should revised sequences cause additional cost.

This subclause is often used in conjunction with Clause 46. The subclause itself only gives the Engineer the power to require the document to be prepared.

It is unlikely that the Contractor will agree that he has been responsible for the delay. The Site Team must act in a manner that protects the Contractor's interests whilst putting into effect the action required by the Engineer.

This can be achieved by writing a letter to the Engineer which summarises all the claims or notices of claims, together with realistic estimates of time entitlement against each. He should state that it is the Contractor's contention that "the Engineer has not properly considered and awarded the time due for these conditions and hence the measures of mitigation taken hereunder will be added to the costs of the individual claims."

It is also possible for the Engineer to order the revision of the programme without operating Clause 46 to see that the works can be completed in extended time. This gives him a document against which to monitor work which may be differently phased as well as delayed.

Subclause (3) is the means by which the Engineer may instruct the provision of detailed method statements and calculations. It should be checked that the working up of tender designs does not lead to unnoticed increases in resources in the method statements above the amounts in the tender. Once declared, the Engineer can assume that the method statement reflects the tender intent, particularly if he has not issued any relevant instructions or directions.

Subclause (4) makes approval in writing necessary and gives the criteria on which method statements may be rejected, which is limited to "fails to meet the requirements of the Drawings or Specification or will be detrimental to the Permanent Works." The Contractor should seek instructions for alterations for any other reason through Clause 13(1).

Subclause (5) makes it mandatory for the Engineer to provide the design information for the Contractor to comply with Subclauses (4) and (5).

Subclause (6) covers two aspects: firstly the possible failure of the Engineer to approve in a reasonable time, and secondly the design information provided causing putative delay and cost.

All submissions under Clause 14 should have a timetable required for approval so that the works are not affected. A standard is then set for what is deemed "reasonable". The Contractor cannot take this as a licence to design the Clause 14 programme so that there is no time for the Engineer to give approval as this is obviously not reasonable.

The effect of a timetable will be beneficial as it concentrates the mind and, if the Engineer does not object to it at submission, he will find it hard to argue that it was unreasonable later on.

If the design criteria supplied under Subclause (5) result in limitations that were not clear in the documents supplied at tender and "could not reasonably have been foreseen by an experienced Contractor" then claims for direct costs and a paid extension can be submitted. Again in this aspect the Site Team will be aware of such limitations if they maintain close contact with, and preferably supervise the work up of the tender methods and require to know the reasons for any increases as stated under the review of Subclause (3). Subclause (7) is just the normal disclaimer leaving responsibility with the Contractor.

Summary Advice for the 5th Edition
Ensure that the submitted Clause 14 programme reflects the tender intent or improvements upon it.
(a) Include a timetable for approvals with all submissions under Clause 14. If the timetable is exceeded give notice of the effects immediately.

(b) If the Engineer rejects proposals the Contractor should ensure, as is mandatory for the Engineer, that the reasons are in writing. Unless they are covered in whole or part by those discussed under Subclause (4) the Contractor should request the Engineer's instructions under Clause 13(1). If this cannot be obtained and revised proposals known to be acceptable have to be made they should have the attached caveat that "the modifications from our previous proposals which matched our obligations under the contract are deemed to be directions under Clause 13(1)."

(c) Do not issue programmes that are not required under the contract unless fault is accepted for delay and a revised sequence is sought against which to judge future events.

6th and 7th Editions: Clause 14

Significant restructuring of the clause has been undertaken to introduce more certainty, a 21 day timetable for programme submissions and approvals and to specifically add profit to any costs.

Subclause (1) is divided into three with the Contractor having 21 days to submit a new programme if it is rejected under Clause 14(2)(b).

Subclause (2) is a new provision that deems the programme accepted within 21 days if the Engineer does not act. Regarding the reasons for rejection in Subclause (2)(b). These are limited to those set out in what is now Subclause (7)(b). These are the same as in the old 14(4) except that Drawings and Specification become "requirements of the Contract" which can bring in other documents within the Contract.

Subclause (2)(c) regarding the provision of further information leads into a new Subclause (3).

Subclause (3) gives the Contractor 21 days to provide the information and the Engineer 21 days to accept the programme or reject under Subclause (2)(b). He is not given the opportunity to ask for further information again. This should then cut off uncertainty and provide an officially accepted programme.

Subclause (4) reflects the old Subclause (2) regarding provision of revised programmes (the same caveats suggested above apply) but introduces the 21 day periods for both the Contractor and Engineer.

Subclause (7) reflects the old Subclause (4) regarding consent, except minor revisions in criteria and the introduction of the 21 day period.

Subclause (8) reflects the old Subclause (6) except that with greater use of Contractor designed elements extra cost of permanent and temporary works is separated and profit is specifically added to cost.

DC: Clause 14

The wording of Clause 14 follows that of the 6th except that it is cut off at Subclause (6). This provides no basis to limit the Employer's Representative's rejection of the programme under Subclause (2)(b). Also there is no provision for recovery of cost if approval of the programme is delayed or design criteria not supplied.

Highways: Clause 14

A Contract Programme is submitted with the tender and is linked to a schedule of payment milestones for what is a lump sum contract. The programme submitted under Clause 14 has less significance.

Clause 14.1 also gives a 21 day (three week) period for submission. It provides for the programming to be a prescribed form which is very prescriptive and is required to be fully resourced. The payment milestones remain as the Contract Programme submitted with the Tender.

Clause 14.3 introduces the Progress Report and its format.

Clause 14.4 enables the Employer's Agent to require a report when delay occurs with reasons for the delay and revised programme to demonstrate completion within the contract or extended period.

FIDIC: Clause 14

Follows the 5th Edition in the first two subclauses. Subclause (3) is replaced with the requirement to produce a cash flow forecast in quarterly periods. The remainder of the subclauses are omitted other than the 5th Edition Subclause (7) disclaiming any responsibility arising from approval, which is now Subclause (4).

The lack of requirement on the Engineer to provide design criteria for the permanent works, the elimination of the need to submit method statements with the programme and the Engineer's duty to say if the methods affect the permanent works, together with the elimination of instruction on mode of construction (in Clause 13) show far less involvement by the Engineer in controlling the technical side of the project.

New Engineering Contract: Clauses 31 and 32

The fundamental difference in the New Engineering Contract towards management of the Contract is that it is to be by the programme. This starts with an Accepted Programme which is altered to deal with the changing circumstances. The succinct clauses cannot be paraphrased and are as follows:

31.1

If a programme is not identified in the Contract Data, the Contractor submits a first programme to the Project Manager for acceptance within the period stated in the Contract Data.

31.2

The Contractor shows on each programme which he submits for acceptance:-

- the starting date, possession dates and Completion Date;
- for each operation, a method statement which identifies the Equipment and other resources which the Contractor plans to use;
- planned Completion;
- the order and timing of:
  - the operations which the Contractor plans to do in order to provide the Works; and
  - the work of the Employer and Others either as stated in the Works Information or as later agreed with them by the Contractor;
- the dates when the Contractor plans to complete work needed to allow the Employer and Others to do their work;
- provisions for:
  - float;
  - time risk allowances;
  - health and safety requirements; and
  - the procedures set out in this Contract;
- the dates when, in order to Provide the Works in accordance with his programme, the Contractor will need:-
  - possession of a part of the Site if later than its possession date;
  - acceptances; and
  - Plant and Materials and other things to be provided by the Employer; and
- other information which the Works Information requires the Contractor to show on a programme submitted for acceptance.

32.1

The Contractor shows on each revised programme:

- the actual progress achieved on each operation and its effect upon the timing of the remaining work;
- the effects of implemented compensation events and of notified early warning matters;
- how the Contractor plans to deal with any delays and to correct notified Defects; and
- any other changes which the Contractor proposes to make to the Accepted Programme.

32.2

The Contractor submits a revised programme to the Project Manger for acceptance:

- within the period for reply after the Project Manager has instructed him to;
- when the Contractor chooses to; and, in any case,
- at no longer interval than the interval stated in the Contract Data from the starting date until Completion of the whole of the works.

The New Engineering Contract contains at Clause 50.3 a heavy sanction in that 25% of interim certificates may be withheld if the Contractor has not submitted a programme for acceptance.

INSURANCE CLAUSES AND EXEMPTIONS LEADING TO PAYMENT

6th, 7th Editions and DC: Care of the Works: Clause 20

These all have the same clause provisions. The Contractor is required to take full responsibility for care of the works, including temporary and permanent materials, plant and equipment, up to the issue of a certificate of substantial completion, Clause 20(1)(a), whereupon this passes to the Employer other than for outstanding works.

The excepted risks, other than war, rebellion and civil disorder, likely to be incurred are:

20 2(a)	use or occupation by the Employer, his servants or agents and other contractors not in the contractors employ.
20 2(b)	fault or error in the design, except where executed by the Contractor
20 2(f)	sonic boom.

The Contractor is required to rectify all damage but is entitled to recover that caused by excepted risks. Clause 20(3)(c) contemplates there being damage caused by two or more events where only one is an excepted risk and that the effects should be split.

FIDIC: Clause 20

Although structured more closely to the 5th Edition, it contains the same provision as the 6th. It has a further exemption, being "forces of nature against which an experienced Contractor could not reasonably have been expected to take precautions".

5th Edition: Clause 20

Essentially, the same provisions except that the liability goes on for 14 days after what was then called a Completion Certificate (still issued under Clause 48), and that damage from an excepted risk as well as a Contractor's risk was not contemplated.

Highways: Clause 21

Similar provisions are contained in 21.1 and 21.1.1 for transfer of the same risk to the Employer once a "Taking over Certificate" for a part or all of the works are issued. Similar provisions are contained in 21.1 and 21.3 regarding rectification and the split of cost if damage occurs from an excepted and Contractor's risk.

The exceptions are more limited in that the war, rebellion and civil disturbance exemptions are limited to occurring in the UK and design errors eliminated. In 21.4.5 the damage from the Employer's occupation is constrained if such occupation was provided for in the Contract and other organisations are not mentioned. The provision in 21.4.6 is comparable with the exception in ICE Clause 21 considered below but is limited to damage to the works which is an inevitable consequence of their execution. **This is only the second clause and the fourth cause that leads to an Employer's Change in the Highways contract without one being instructed for different works.**

New Engineering Contract
The definitions are in Clause 80.1, the concept in Clause 80.3 and the period in Clause 81.1, which is where the difference is in that insurance is to the Defects Correction Certificate and not the Completion Certificate.

INSURANCE OF THE WORKS

6th, 7th Editions and DC
The sum to be insured is the full value of the works and things for incorporation plus 10% to cover demolition.

FIDIC: Clause 21
Follows 6th and 7th Editions but the reinstatement cost is to include profit and 15% is to be included for fees and demolition.

5th Edition: Clause 21
Only the value of the work needed to be insured together with the Contractor's constructional plant.

Highways: Clause 22
As 5th Edition but sum insured to cover inflation and Employer's fees.

New Engineering Contract: Clause 22
The required insurances for any project are set out in the Insurance Table in Clause 84.2. Any required additions to cost of replacement will be in the Contract Data.

DAMAGE TO PERSONS AND PROPERTY

5th Edition: Clause 22
This clause effectively covers damage to persons (other than employees under Clause 25) and property other than the works, both within and without the site boundary. This is then mitigated by the extent of any contributory effect by the Employer, his servants or agents through Subclause (1)(a). There then comes the list of exceptions in Subclause (b)(i) to (v) for which Subclause (2) provides a

cross indemnity by the Employer. The most significant of these is item (iv) "damage which is the unavoidable result of the construction of the works in accordance with the contract."

The reason for this inclusion in the contract is that insurance is only available for accidental damage and not for something which is inevitable or almost so. A classic example is the effects of driven piling in a built-up area. This exception also covers damage caused by unforeseen ground conditions if these cause damage to the works constructed at the time which would have been unavoidable had the conditions actually discovered being applied to the design. For a case in this category the Contractor would receive time and consequential costs under the Contract.

6th, 7th Editions, DC and FIDIC: Clause 22
Although differently structured, the provisions are essentially the same.

Highways: Clause 21
The similar provision within 21.4.6 is limited to the works and so the Contractor must insure all other risks.

New Engineering Contract: Clause 80.2
The provisions are very similar to the ICE Contracts but worded differently. These are in Clause 80.2 where from starting date to the Defects Certificate the Employer bears risks due to occupation of the site by the works or for the purpose of the works which is the unavoidable result of the works, negligence by anyone connected to the Employer, damage to the works by fault of the Engineer or his design, war and radioactive contamination and any other Employer risks in the Contract Data.

ACCIDENT AND INJURY TO WORKMEN

5th, 6th, 7th Editions, DC, Highways and FIDIC: Clause 24 [Highways 25]
This clause is designed to cut off the Employer from any liability for the Contractor's employees and ensure that the Contractor has in force the appropriate Employee Liability Insurance.

New Engineering Contract
Defined as a Contractor's risk in Clause 81.1 and in the Insurance Table.

EVIDENCE AND TERMS OF INSURANCE

5th, 6th, 7th Editions, DC and FIDIC: Clause 24
[Highways 24 and 27] NEC Clause 86.1

Provides the capability for the Employer to insure if the Contractor fails to do so.

NOTICES AND FEES

6th, 7th Editions and DC: Clause 26(1) and (2)
Provide for the Contractor to give all notices and pay all fees (6th includes Street Works Act by Corrigenda and DC in 26(3)(b)) required but allows for them to be recovered through the Contract.

5th Edition
Excludes the notices for the Street Works Act but where this Contract is used in the UK it is likely to be modified.

FIDIC
Contractor to give all notices and pay all fees; no automatic right of recovery.

Highways: Clause 32.2
Contractor gives all the notices and pays the fees without recovery.

New Engineering Contract
No specific provisions in core clauses.

STATUTES

5th, 6th Editions and DC: Clause 26(3)
All statutes are to be complied with. Additional payment is available through variations under Clause 51 (5th, 6th and DC) and 53 (7th) if:

- a) the construction of the works results in an unavoidable breach of statutory regulations; or
- b) an instruction is in breach.

Obtaining planning permission is also the Employer's responsibility.

7th Edition
Similar provisions to 5th and 6th Editions except that planning permission is limited to the permanent works and the temporary works in their final position.

FIDIC: Clause 26
Requires compliance with National, State or Local law and rules and regulations of public bodies and companies whose rights are affected. It does not have the specific remedies in the 5th but a similar result can be obtained through legal impossibility under Clause 13. Any permissions to allow the works to commence remain with the Employer.

Highways: Clause 32

Similar provisions to 5th and 6th except that it also specifically provides for changes in statutes that result in a change to the design probably because tax changes are the Contractor's risk.

The Highways also introduce at 32.3 compliance with the Special Requirements of Statutory Bodies (normally in Clause 72 or higher numbers on the ICE contracts). However, these bodies must comply with the appropriate Special Requirements in the documents and the appropriate legislation. They often don't and an Employer's Change is necessary to alter the documents to include non standard requirements.

The first clause potentially contains three causes for entitlement to a change under the Highways conditions without an instruction. Theses can be summarised as illegality, statutory change and a named body not complying with the appropriate regulations.

New Engineering Contract

No specific provisions in Core Clauses.

STREET WORKS ACT

5th, 6th and 7th Editions: Clause 27

The 5th and 6th (in the original issue) referred to the 1950s. The latter was corrected by corrigenda to recognise the 1991 Street Works Act. Both contracts will almost certainly be amended to include the provisions in the 7th Edition. The Employer is to obtain the Street Works license and the Contractor to give notice of commencement. Conditions or restrictions to the licence post award lead to variations and variations which involve further notices and statutory periods will attract time and costs.

DC and Highways

In the former Clause 27 is eliminated but at 26(3)(d) the Contractor is relieved from obtaining the licences unless specifically stated otherwise in the Contract. The latter makes no specific reference to the Street Works Act but there are specific requirements at 30.3 regarding traffic management and the responsibility for licences and costs are with the Contractor through Clause 32.2.

FIDIC

No specific provisions but similar duties through Clause 26 reviewed above.

New Engineering Contract

No specific provisions in Core Clauses.

MATERIALS, WORKMANSHIP AND TESTS

5th, 6th, 7th Editions: Clause 36

Subclause (1) provides for all materials to be in accordance with the Contract. It is made mandatory for the Contractor to supply all materials for testing before their inclusion in the works, together with the means of testing and to carry out the tests.

Subclause (2) makes the costs of samples for tests prescribed under the contract to be borne by the Contractor and for any other tests by the Employer. There is no reference to the subclause to the results of the tests so that payment for the test samples would appear to be independent of the results of the test.

Subclause (3) allows payments for tests when they were not included in the documents, or for an incompletely itemised test of fitness for purpose of the completed structure, provided in both cases that the results are satisfactory. The Clause, as far as testing is concerned, is in many ways similar to the provisions of Clause 50, but in both cases the Contractor is paid for the cost of unprescribed tests only if they have satisfactory results. It is said that it was the Drafting Committee's intention that Clause 50 be invoked in the Maintenance Period and 36 prior to that. The Engineer can use either with similar effect.

FIDIC: Clause 36

Follows 5th except that Subclause 3 is divided with tests not prescribed being in Clause 36.4. Clause 36.5 introduces a right to Extension of Time for successful non prescribed tests.

DC: Clause 36

As 5th, 6th and 7th, except that at Subclause (6) ordered variations lead to the Contractor considering whether further tests are necessary.

Highways: Clauses 38, 39 and 52

Clauses 38 and 39 broadly follows provision of ICE Clause 36 but more prescriptive. Clause 52 mirroring Clause 50 of the ICE is titled "Search for Defects" and specifically is for the maintenance period. Successful tests or negative searches result in payment.

New Engineering Contract: Clauses 40 to 42

Clause 40 requires that tests required by the Contract be carried out in a manner which does not delay the works. Repeated tests due to defects are assessed for cost by the Project Manager.

Clause 41 requires that no materials be brought to site before requisite tests are done.

Clause 42 permits the Project Manager to order searches for defects; however, this is linked to payment through Clause 60.1(10).

EXAMINATION OF WORK

5th, 6th, 7th Editions, DC and FIDIC: Clause 38
Provides that the Engineer must be given the opportunity of inspecting all work before it is covered over.

   If this opportunity is not afforded to him the Contractor will not be paid for making holes or openings through to the work that was covered up, whether it is found to be satisfactory or not.

Highways: Clause 40
Provisions are in Clause 40 and are similar but include the Designer who works for the Contractor but has to certify the work.

New Engineering Contract
See comments on Clauses 40 to 42 above.

REMOVAL OF IMPROPER WORK

5th, 6th, 7th Editions and FIDIC: Clause 39
The Engineer is provided with the power to instruct the removal of materials with full rectification. If the Contractor does not comply the Engineer can have the work executed by others at the Contractor's expense.

   Subclause (3) allows the Engineer to reverse previous approvals by himself or the Engineer's Representative. It is not mandatory for the Engineer to have tests carried out under either Clause 36 or 50 before ordering removal. If this happens when the suitability of the work or materials is disputed and the consequences to the Contractor are great, the site team must steer the Engineer or his Representative into testing. Facts are then established which can be debated and an avenue has been opened for recovery of time and cost. If the work is clearly defective the Contractor should not need to wait for a Clause 39 notice to honour his obligations.

DC: Clause 39
As 5th, 6th and 7th, except that the Employer does not have the right to get others to remove defective work contained in 39(2) above.

Highways: Clause 41
Similar to the ICE except that the Contractor must act in 14 days or the matter becomes a dispute and goes to Adjudication.

New Engineering Contract: Clause 43
Requires that the Contractor correct all defects whether or not notified to him within the Defects Correction Period and the Project Manager is to arrange access.

The New Engineering Contract then goes on to prescribe for two eventualities which regularly occur and are not readily dealt with in the other contracts. Within Clause 44 is the provision for acceptance of a defect based on a quotation from the Contractor leading to a reduction in the prices or an earlier completion. The other is in Clause 45 where the Project Manager may assess the cost of having a defect corrected by others and the Contractor pays this amount. (The work does not have to be done.)

SUSPENSION

5th Edition: Clause 40: Suspension of Work
Subclause (1) deals with the powers of the Engineer to suspend all or part of the work with the provision for payment.

The Contractor should never stop work on any section of the job, however obvious the need, unless a suspension order is given which triggers the mechanism of this clause linked to Clauses 52(a), 44 and 60 for payment.

Subclause (2) allows the Contractor, after a suspension notice has been in force for three months, to write to the Engineer requesting to be allowed to proceed with the suspended work.

If this is not forthcoming the suspended area can be treated as an omission under Clause 51. If this applies to the whole of the contract it is considered to be an abandonment of the contract by the Employer and the Contractor should visit the lawyers!

6th, 7th Editions and DC: Clause 40
As 5th but profit has been specifically added to cost.

Highways: Clause 42
Broadly follow the 6th, 7th and DC; the fourth item to automatically yield an Employer's Change.

FIDIC: Clause 40
Basically follows the 5th but the Employer is to be consulted before time or an additional "amount" is to be added to the Contract Price. This is a greater risk than standard ICE Contracts.

New Engineering Contract: Clause 34
The Project Manager may instruct the Contractor to start, stop and/or restart the works. This is a Compensation Event (see under Variations below). Clause 95.6 permits the Contractor to terminate the Contract after 13 weeks if the Project Manager has not ordered a restart.

DATE FOR COMMENCEMENT

6th, 7th Editions and DC: Clause 41: Commencement
Provide a range of three choices being either the date in the appendix to the Form of Tender, a date instructed by the Engineer within 28 days of award (DC between 14 and 28 days), or such other agreed date.

5th Edition and FIDIC: Clause 41
Provides only the latter option and as soon as reasonably possible.

Highways: Clause 43
Site commencement date or dates with provisions for phased possession are dates before which works cannot commence. Date(s) are likely to be included in the Contract documentation.

New Engineering Contract: Clause 30.1
The Contractor does not start until the first possession date which will be in the Works Information or Contract Data.

POSSESSION OF SITE

5th Edition: Clause 42: Possession of Site
There are two significant points in this clause. Firstly, that payment flows from the inability to allow the Contractor to commence.

Secondly, that the portions of the site have to be released to match progress anticipated on the Clause 14 programme once this has been approved and providing it matches any phased release in the contract documents.

The Contractor is not entitled to automatic full release of the site. This again reinforces the need for clarity in the Clause 14 programme and method statements.

Subclause (2) makes it necessary for the Contractor to obtain all way leaves and any additional land that may be required beyond the site boundary.

6th Edition and DC: Clause 42
Clause 42(1) of the 5th is broken up into new Subclauses (1) to (3). The significant change is that provisions for access to all or part of the works are added in Subclauses (1)(c) and (2)(b) and that, in addition to delay costs, any additional temporary or permanent works costs to which profit is now specifically added is associated with varied access.

7th Edition: Clause 42
Subclause (2)(a) is modified to require that the Contractor be given possession of the whole of the site, but access to all of it is not mandatory other than when required.

FIDIC: Clause 42

Anticipates that Part II may provide information on phased release of the site and, again, links access to that necessary to conform to the Clause 14 programme. Clause 42.2 leads to time and cost on failure to afford possession as defined above and similar provisions for way leaves are now in Clause 42.3.

Highways: Clause 44

Follows the provisions of the 5th in a different format. The difference is that possession may be withheld if insurances are not in place and in such a case there would be no extension granted.

Clause 44.3.1 and 2 lead to extension of time and sums due based on cost.

New Engineering Contract: Clause 33

Possession of each part of the site is to be given to the Contractor on or before the date shown on the Accepted Programme.

Clause 33.2 envisages that the Contractor may have to provide facilities for the Employer and includes provision for the Project Manager to determine sums to be paid to the Employer if the Contractor fails to do so.

TIME AND EXTENSIONS

5th Edition: Clause 44: Extension of Time

Whilst studying this clause it should be recognised that the award of time under Clause 44 carries no financial reward. It is a device whereby the contract period is modified and liability for liquidated damages mitigated. The only way that money can be recovered by virtue of extension of time being granted is by being able to claim under some other clause. It must be recognised that there is no provision in the contract for the recovery of overheads on its own. The Contractor must be able to prove realistically that delay arises from variations or, in certain cases, disruption to programme beyond his control.

Subclause (1) introduces the point in the contract where exceptional weather conditions may be considered and hence, as stated in the introduction to this clause, there is no payment for this. "Full and detailed" particulars are required within 28 days or "as soon thereafter as is reasonable in all the circumstances" but for the important reason which comes at the end of this subclause "in order that such claim may be investigated at the time".

Subclause (2) requires the Engineer to make an interim award of an extension of time if such is warranted and if not he must tell the Contractor.

Subclause (3) envisages that the Engineer may award an extension of time when one has not been asked for. Some Engineers have used this power to award extensions for exceptionally adverse weather for greater periods and in parallel to causes where the evaluation under Clauses 52 and 60 might include for establishment costs. If this is likely and the Contractor does not require an extension for weather effects he should state how he is coping with the weather and minimising any obvious problems. The subclause lays a duty upon the Engineer to take into account all circumstances known to him when making a final award, whether notified or not. The Contractor must not rely on this because the Engineer may well not be aware, or claim not to be aware, of the effect or extent of any delay.

Subclause (4) requires the Engineer upon issue of the Completion Certificate, to review all circumstances and finally determine the overall extension of time warranted. However, he cannot reduce awards already made.

6th Edition and DC: Clause 44

The 5th Edition Subclause (2) is divided into Subclauses (2) and (3). 5th Edition Subclauses 44(2) and (3) become (4) and (5) and are altered to the extent that a 14 day time limit is introduced.

New Subclauses (2) and (3) deal with interim applications and awards. The particular practical changes are within Subclause (3) where the interim award is to be made forthwith and only if extension will be actually required for completion. This only serves to emphasise the difference between time-related cost and extension of time.

The assessment of the due date for completion and final determination are now to be executed in 14 days.

7th Edition: Clause 44

As 6th but with two new Subclauses (1)(e) and (f) which cover Employer Default. The final determination in Subclause (5) is specifically to be reasonable. There is no practical difference.

Highways: Clause 46

Covered in Clause 46 and entitled "Variation in Period for Completion" and in 47 "Failure to Provide Particulars".

Taking the latter first, the major difference is that the Contractor has to apply for extensions and has a prescriptive timetable of 21 days after the event (47.1.1) for notification and 21 days after the notification for full particulars (47.1.2) to be entitled to time. There is no caveat on this as in Highways Clause 55.4 regarding claims for changes in the Contract Sum.

Within Clause 46.1.8 is a very prescriptive list of matters which are to be included within the notice. The most onerous are:

1.   Details of the consequences whether direct or indirect upon completion of sections or the whole works.

2.   Details of proposed mitigating measures and an estimate of reasonable cost.

The examination of the list of events entitling the Contractor to an extension of time in Clause 46.1 to 46.8 shows that exceptional weather and, hence, its associated effects as well, are not included. These matters are the Contractor's risk and he has, therefore, to accelerate at his own cost to avoid L & ADs.

Similar provisions to those in the ICE regarding review on completion are in 46.3. Clause 46.4 recognises that agreement of Employer's Change can include agreement of additional time without the review process in this clause.

Finally, 46.5 rules out claims for finance costs for later payment of milestones through these events.   It is clear that the overall requirements are far more onerous and the site team have far more to do at the time.

FIDIC: Clause 44
Follows the 5th in terms of the events leading to an extension of time and then adds Employer default in line with the 7th. The Engineer is only bound to make a determination if the Contractor notifies him of the event within 28 days of it first arising and then provides full particulars 28 days after that notification. (There are exceptions if the effects are ongoing at the end of 28 days and are not fully determinable.)

New Engineering Contract
The issue of time is bound up in the assessment of Compensation Events, Clause 63, which is considered within "Valuations of Variations" below. The timescales for action by the Project Manager on notification by the Contractor is two weeks.

The procedure described in Clause 63.3 is based on changes to planned completion dates and therefore avoids the question of whether an extension is required to avoid L & ADs.

SUNDAY AND NIGHT WORK

5th, 6th, 7th Editions and DC: Clause 45: Sunday and Night Work
The provisions of this clause for written permission to work at night and on Sunday are not onerous, given the exceptions of the second sentence. However, often strict requirements are laid down in additional clauses which, through the wording of this clause, would take precedence.

FIDIC: Clause 45
As 5th but includes caveat that work that is usually to be carried out by multiple shifts is exempt. This is an area where Part II will contain specific requirements.

Highways
Contains no restrictions but those from noise in Clauses 30.1 and 30.4.

New Engineering Contract
No specific provisions in conditions.

CONTRACTOR DEFAULT REGARDING PROGRESS

5th Edition: Clause 46

The effect of the clause is clear; if the Contractor is behind for reasons which are obviously his fault then he can be required to complete by the notified total or sectional completion date at his own cost.

Some Engineers apply this clause to back up their stand on contentious issues where the Contractor's application for an extension is being challenged. If the Contractor obeys instructions under this clause without comment he could be deemed to have accepted responsibility for the delay.

> **Summary Advice**
> Unless fault is accepted the Contractor should always respond in the following manner: "because of our submissions of...which include applications for extension of time that you have not properly considered, this instruction is not in our opinion valid under Clause 46.
> We will execute your instructions but without prejudice to our further submission of a claim under Clause 13(3) to be valued in accordance with Clauses 52(4) and 60."

FIDIC

As 5th Edition, but provides for the Employer to offset any additional supervision costs.

6th, 7th Editions and DC: Clause 46

New Subclause (2) splits reference to permission to work nights and Sundays and adds the caveat that this is on site.

Highways: Clause 48

Provisions are in Clause 48 "Rate of Progress" and largely follow the 5th except that there are provisions to recover any costs incurred by the Employer in "monitoring design execution or construction of the works".

New Engineering Contract

No specific provisions but this will be controlled through the periodic required revisions to the Accepted Programme (see Contract Programmes earlier).

ACCELERATION

5th Edition

The Engineer is unable to order acceleration for the reasons set out under "Variations".

6th, 7th Editions and DC: Clause 46

Subclause (3) brings in the provision for special terms to be agreed for acceleration, if required by the Employer when the Contractor is not in default, before the measures are implemented. This regularises the position described later under Clause 51 for the 5th Edition.

Highways: Clause 55

Theoretically within the Employer's Agent's powers as he may under Clause 55 instruct changes to the Contractor's "obligations". (See under "Variations" below.)

New Engineering Contract: Clause 36

This permits the Project Manager to instruct the Contractor to submit a quotation for an acceleration to achieve completion before the due date. (This is usually the original date with the due date having been moved by Compensation Events.)

The quotation changes the prices and the Completion Dates, as well as a revised programme.

LIQUIDATED DAMAGES

5th Edition: Clause 47: Liquidated Damages

Introduces into the contract the dates or sectional dates which are in the Appendix to the Form of Tender. The Contractor's duty is to complete the work to time. If he is prevented from so doing, he must give all necessary notices under the various clauses linked to Clause 44, so that an extension of time may be obtained. The required procedure is covered under the analysis of Clause 44 in the next section. Should Liquidated Damages have been deducted from an interim certificate then it is likely that either the Employer or the Engineer are in breach of the two preconditions for deduction.

These are that the Engineer: "...shall under Clause 44 (3) or (4) have determined and certified any extension of time to which he considers the Contractor entitled" and "shall have notified the Employer and the Contractor that he is of the opinion that the Contractor is not entitled to any further extension."

FIDIC: Clause 47
Follows the 5th Edition but refers only to Clause 43 and this clause provides the link to extensions that have been granted under Clause 44.

6th, 7th Editions and DC: Clause 47
The new Clause 47(6) considers what happens if, after liquidated damages have become payable in respect of part of the Works, the Engineer issues a variation order or adverse physical conditions occur or any other item warranting an extension.

Effectively L & ADs are suspended. This position was developed in Case Law and was then included in the ICE conditions. There is therefore practically no difference between the contracts.

Highways: Clause 49
Provisions in Clause 49 are similar to the ICE 5th.

New Engineering Contract
Liquidated Damages provisions are in Secondary Option R and there are no provisions within the core clauses. If Option R is not used the Employer can sue for damages at law for whatever actual costs he can demonstrate.

COMPLETION CERTIFICATES

5th Edition: Clause 48: Completion Certificate
Completion under the ICE conditions does not mean either totally complete or defect free. If there is a dispute as to whether the Works are complete the test is whether it is "de minimus" and would not affect use or occupation.

The Completion Certificate brings several benefits:

- release of the first part of the retention
- release, after 14 days, of the need to insure for the duties involved in "care of the works"
- if issued before the time for completion or extended time, removes the threat of liquidated damages or, terminates recovery of liquidated damages if these are being recovered.

Subclause (1) makes it the Contractor's duty to write to the Engineer requesting a Completion Certificate and the Engineer to respond within 21 days with the Certificate or with a list of items to be completed before one is issued. The Contractor must have undertaken in writing to execute any outstanding work.

Subclause (2) dealing with completion of sections of the work makes it mandatory for the Engineer to issue a Certificate of Completion, subject to his satisfaction, where (a) there was a separate date in the Appendix to the Form of Tender, and (b) where the part is occupied or used by the Employer.

Subclause (3) allows the Engineer to use his discretion in considering other parts of the works. It is obviously better to apply where possible under Subclause (2).

6th and 7th Editions: Clause 48

Basically as the 5th with premature use by the Employer entitling the Contractor to a Completion Certificate as a separate item under Subclause (3).

DC: Clause 48

As 6th and 7th with further requirement of operational instructions being handed over before the Completion Certificate. Part of this would, of course, be necessary under CDM in the UK and would constitute the Health & Safety File.

Highways: Clause 50

Termed a "Taking over Certificate" in Clause 50; otherwise, broadly follows the 5th Edition but with the following amendments:

a) the Designer (the Contractor's man) has certified the design;
b) contract-specific taking over procedures are followed;
c) a written undertaking is given to complete outstanding works in six months.

FIDIC: Clause 48

Basically as 5th Edition, but uses the term "Taking Over Certificate". The Contract states in Clause 61 that the works are not complete until the Maintenance Certificate is issued.

New Engineering Contract: Clauses 13, 30 and 35

Greater clarity is provided as to what constitutes completion, which can be defined differently on every contract by the Employer. The details are within the definitions at Clause 13:

(13) Completion is the date, decided by the Project Manager, when the Contractor has done all the work which the Works Information states he is to do by the Completion Date and has corrected notified Defects which would have prevented the Employer using the works.

Thus, the precise terms can be different on every contract, being part of the Works Information. Clause 30.2 provides that the Project Manager certifies completion within one week of it being achieved. It is not dependent upon the Contractor requesting a certificate. Clause 35 deals with Taking Over by the Employer. He must do if he occupies all or a section.

He need not take over immediately if the Contract Date states he is unwilling to do so, but must do within two weeks which provides a back stop for insurance liability.

Unlike the ICE-based contracts, insurance continues to the receipt of the Defects Certificate.

MAINTENANCE AND DEFECTS

5th Edition: Clause 49: Maintenance and Defects
Subclause (1) defines the Period of Maintenance.

Subclause (2) provides that the Contractor should finish all works outstanding at the issue of the Clause 48 Certificate of Completion as soon as possible.

The Engineer is required to write to the Contractor at the end of the maintenance period, or within 14 days thereafter, with details of repairs required. It should be noted that fair wear and tear is excepted and this can often lead to argument.

Subclauses (3) and (4) provide the means of determining responsibility for cost and the remedy on the Contractor's failure to execute the work.

Subclause (5) introduces the obligation on the Contractor to execute all necessary repairs to highways immediately with or without instructions.

FIDIC: Clause 49
As 5th but without provisions in Subclause 5.

6th, 7th Editions and DC: Clause 49
As 5th except that provisions regarding temporary reinstatement in Subclause (5) have been omitted and Subclause (1) has been amended to permit agreements to complete the outstanding work in a lesser period than is now named the Defects Correction Period.

Highways: Clause 51
Confirms in 51.2.1 the need to complete in less than six months, otherwise basically similar to the ICE except at 51.7 which reintroduces measures and approvals necessary for Traffic Management.

New Engineering Contract: Clauses 43, 44 and 45
Requires that the Contractor correct all defects whether or not notified to him within the Defects Correction Period and the Project Manager is to arrange access.

The New Engineering Contract then goes on to prescribe for two eventualities which regularly occur and are not readily dealt with in the other contracts.

Within Clause 44 is the provision for acceptance of a defect based on a quotation from the Contractor leading to a reduction in the prices or an earlier completion. The other is in Clause 45 where the Project Manager may assess the cost of having a defect corrected by other and the Contractor pays this amount. (The work does not have to be done and in effect it is a penalty for not complying with the contract unless the Employer has something of a lesser value.)

VARIATIONS

5th Edition: Clause 51: Ordered Variations

Subclause (1) gives the Engineer the power to order variations to the works that are "necessary for the completion of the works" and for "satisfactory completion and functioning of the works". From this it follows that, given the necessary conditions for a Clause 48 Certificate, variations cannot be ordered after the issue of a Clause 48 Certificate.

It is held that the Engineer cannot under the Conditions of Contract order acceleration. Subclause (1) refers to variations in the "specified sequence" and "timing", but these are those established in the Clause 14 programme and method statements. Clause 13(2) introduces "the mode, manner and speed of construction.... conducted in a manner approved by the Engineer".

Clause 14(2) introduces the only situation where there is a recognised power to order a change in the programme. It has to be "necessary to ensure completion of the work or any section within the time for completion as defined in Clause 43 or extended time as granted pursuant to Clause 44(2)". Hence, there are no powers to order acceleration other than to regain programme as instructed under Clause 46 (see under this clause). It is held that if the Employer still requires acceleration this can only be done by it being priced as an extra to the contract and not under the contract.

Subclause (2) makes it imperative for the Contractor to confirm quickly all oral instructions either by use of a standard format or by letter, either to the Engineer or Engineer's Representative, depending on the provisions of Clause 2.

Subclause (3) identifies changes in quantities are a variation although not requiring an order in writing and therefore extension of time can flow.

Summary Advice

If acceleration is ordered by the Engineer, the Contractor must carefully consider his position. Legal advice should be sought. The Contractor should eventually send the following as an absolute minimum:

"As acceleration cannot be ordered under the ICE 5th Edition we give our price on an extra contractual basis for attempting to comply with the requirements notified to us."

FIDIC: Clause 51

Subclauses (1) and (2) basically follow the 5th but do not contain the requirement for instructions to be in writing or specify a procedure for confirmation. It would be unwise not to do so. Again, acceleration is excluded for the reasons discussed under the 5th and there is a provision for variations without payment where the cause is the Contractor's fault.

6th and 7th Editions: Clause 51

Subclause (1) is divided into (a) and (b) and now specifically includes for variations in the Defects Correction Period.

Subclause (3) is added with a specific caveat on variations caused by Contractor default.

Acceleration can now be ordered by agreement through the Clause 46 modifications as examined earlier under "Acceleration".

DC

Follows the 6th and 7th but the Engineer's Representative has powers to alter only the Employer's Requirements and not methods, sequences or timing. Alterations to these that result from such a variation will be paid.

Highways

The variation clauses are divided into three parts in common with the 7th and DC, being 53 Changes; 54 Valuation of Employer's Changes; and 55 Procedure for Claims. However, Clause 53 also provides for a procedure whereby Employer's Changes are priced and agreed which then obviates the need for the second two clauses. There are very specific definitions of what constitutes a variation for which the Contractor will be paid.

53.1 Change: the definition covers the same ground as the 5th, 6th and 7th, even envisaging change in the Contractor's Proposals and variations in sequences and method and an additional catch-all of variations in the Contractor's obligations. What initially seems a wide, liberal approach is actually designed to ensure that all alterations to the Contractor's proposals are put to full scrutiny.

53.2 Employer's Change: effectively the Employer has to require and instruct something different. When he does so it may result in a change to the Contract Sum and/or extension of time. There are a few clauses examined earlier which require the Employer to issue an Employer's Change and they would fall within the change of Contractor's obligations.

The true limitation of what the Contractor will be paid for is contained in the further subclauses which define what are Contractor's changes of various types.

53.3 Contractor's Changes: these are variations to be requested in writing by the Contractor and agreed by the Employer's Agent. Consequently, the

Contractor cannot change his proposals in any way without going through the procedure outlined below.

53.3.1 A Contractor's Change is here defined as where the Employer's Agent considers that the Contractor's Proposals need amending to comply with his obligations (effectively because Employer's Requirements take precedence) or is satisfied that there is no change in the quality or standard of the works. Through 53.3.4 there is no change to the Contract Sum.

53.3.2 A Contractor's minor change permits a minor change at the Employer's Agent's discretion which contemplates a reduction in standard or quality which does not affect the whole. Again, through 53.3.4 there is no change to the Contract Sum.

53.3.3 A Contractor's Change resulting in a saving recognises that there will be a reduction in standard or quality and the Employer's Agent is to be satisfied that the saving warrants the change. The negotiation of the size of these savings is stacked in the Employer's Agent's favour as he does not have to proceed even when offered a good deal. The Change cannot alter the Contractor's obligation to perform the Contract.

53.4 Confirmation of other Changes provides in the first half of the clause the procedure for confirming Employer's and Contractor's Changes with endorsed certificates before commencement. The second part of the clause provides for oral agreement of Contractor's Changes only, i.e. those which do not include any reduction in quality or standard, which have to be confirmed in writing as soon as possible. If this is within a seven day period it gives protection from the final provision which only gives the Employer's Agent 14 days thereafter to contradict before it is deemed to have been incorporated as a Contractor's Change.

53.5 Confirmation of Contractor's Minor Change: follows the provisions of 53.4 including oral confirmation.

53.6 Confirmation of Contractor's Change resulting in a saving is restricted to countersigned certificates with no provision for oral agreement.

53.7 Covers the information the Contractor must supply to the Employer's Agent on request regarding potential Employer's Changes, being:

Sub-subclause (1)	the value plus supporting information based on the values in the payment schedule;
Sub-subclause (2)	the length of EoT considered appropriate under Cl 46. In giving this estimate the Contractor must bear in mind the length of the procedure in 53.8 and 53.9;
Sub-subclause (3)	adjustments to milestone payments;
Sub-subclause (4)	any effects on design warranties.

53.8 outlines the provisions when agreement is reached, which is not effective until it is countersigned and covers all the matters in 53.7.

53.9 provides essential curtailment of uncertainty in the position when the estimates cannot be agreed, namely:

Sub-subclause (1)   the Employer's Agent issues the instruction any way and valuation is in accordance with Clause 54:

Or

Sub-subclause (2)   he withdraws the request and under Clause 53.10 the Contractor has no recourse to any extra cost.

The New Engineering Contract: Clause 60

The ICE-based contracts recognise types of variations in specific clauses and then give the Engineer the ability to instruct on any matter. The New Engineering Contract draws all matters together in a single clause. The details of what constitutes a Variation (a Compensation Event in this contract) is contained in Cause 60 and is reproduced in its entirety as it cannot be paraphrased easily.

(1)   The Project Manager gives an instruction changing the Works Information or the boundaries of the site except:

- a change made in order to accept a Defect;
- a change to the Works Information provided by the Contractor for his design which is made at his request or to comply with other Works Information provided by the Employer;
- a change made resolving an ambiguity or inconsistency which the Contractor failed to notify to the Project Manager as soon as the Contractor was aware or should reasonably have been aware of the same;
- a change arising from a failure by the Contractor to reply to a Communication within the period for reply pursuant to Clause 13.3; or a change made to the Works Information as a result of a proposal by the Contractor pursuant to Clause 53.8.

(2)   The Employer does not give possession of a part of the Site by the later of its possession date and the date required by the Accepted Programme.

(3)   The Employer does not provide something which he is to provide by the date for providing it required by the Accepted Programme.

(4)   The Project Manager gives an instruction to stop or not to start any work.

(5)   The Employer or Others do not work within the times shown on the Accepted Programme or do not work within the conditions stated in the Works Information.

(6)   The Project Manager does not reply to a Communication from the Contractor within the period required by this Contract.

(7)   The Project Manager gives an instruction for dealing with an object of value or of historical or other interest found within the Site.

(8)   The Project Manager changes a decision which he has previously communicated to the Contractor.

(9)   The Project Manager withholds an acceptance (other than acceptance of a quotation for acceleration or for not correcting a Defect) for a reason not stated in this Contract.

(10)   The Project Manager instructs the Contractor to search and no Defect is found unless the search is needed only because the Contractor failed to give notice of, or gave insufficient notice of doing work obstructing, a required test or inspection.

(11)   A test or inspection done by the Project Manager causes unnecessary delay.

(12)   The Contractor encounters conditions which:

- are within the Site;
- are not weather conditions or attributable to weather conditions;
- are not naturally occurring surface or sub-surface ground or water conditions or attributable to such conditions; and an experienced Contractor would have judged at the Contract Date to have such a small chance of occurring that it would have been unreasonable for him to have allowed for them.

(13)   A weather measurement is recorded:

- within a calendar month;
- before the Completion Date for the whole of the works; and
- at the place stated in the Contract Data;
- the value of which, by comparison with the weather data, is shown to occur on average less frequently than once in ten years.

(14)   An Employer's risk event occurs.

(15)   The Project Manager certifies takeover of a part of the works before both Completion and the Completion Date.

(16)   The Employer does not provide materials, facilities and samples for tests as stated in the Works Information.

(17)   The Project Manager notifies a correction to an assumption about the nature of a compensation event.

(18)   A breach of contract by the Employer which is not one of the other compensation events in this Contract save to the extent it is caused or contributed to by the Contractor or any Subcontractor.

The main difference to the ICE Conditions is that there are no method-based variations (provided, of course, that there are not specific requirements in the Works Information that the Contractor is complying with.) The other difference, which is an improvement for Contractors, is in (6) where the Project Manager has a set period for response to a communication.

VALUATION OF VARIATIONS

5th Edition: Clause 52: Valuation of Variations 52(1) to (3)

Subclause (1) introduces consultation with the hope of agreement but presupposes valuation at the Contract rates, failing agreement.

Subclause (2) introduces analogous rates. Contractually, the Contractor can be forced to do additional work at a loss. There must always be a reason for rerating of the work, other than that of a loss being made.

If the Contractor is making a loss on "relevant rates" his only course is to concentrate on investigating and demonstrating any differences of timing, rate of working, numbers or uses of temporary works items and the like.

This latter category will be most often connected with reduced quantities and Clause 56(2). The tender will govern whether the Contractor's efforts are directed towards a rates-based exercise or convincing the Engineer that a separate "fair valuation" based on the recorded cost of executing the works should be made.

Subclause (3) brings in the provision for dayworks. The central point here is that the Contractor has no inherent right to have any work executed at daywork rates unless they have been so ordered in writing before its execution.

It has been contended by some Engineers that, where they have included a provisional sum for dayworks, they are allowed to spend up to that sum on any item without the Contractor being entitled to an extension of time. This is manifestly unreasonable, and is certainly not so. The effect as well as the content of any addition or variation to the "Works" must be evaluated (Clause 51(1)). It is not reasonable to say that an abstract provision for expenditure is part of the works.

However, it would also be impossible to argue that daywork under the total of the provisional sum, but not on the critical path, was cause for an extension.

6th Edition: Clause 52

Follows broadly the provisions of the 5th Edition, except that Bill of Quantities rates are now unlikely to apply to variations in the Defects Liability period through the modifications to Subclause (1). The daywork procedures are transferred to Clause 56 "Measurement" and Subclause (4) and are simplified.

7th: Clause 52

Clause 52 has been rewritten to encourage agreement of prices and time effects.

Subclause (1) provides that the Engineer may request, and the Contractor then shall supply, a quotation including estimates of delay and consequential costs. It is recommended that agreement be reached before the order is issued or work starts, but this is not a condition precedent and, in any case, is in the Engineer's hands.

Subclause (2) applies where agreement has not been reached or the Engineer did not request a quotation. After receipt of the variation the Contractor:

Sub-subclause (a) shall submit

(i)	his quote, with due regard to the rates and prices
(ii)	his estimate of delay
(iii)	his estimate of the cost of delay

Within 14 days the Engineer:

Sub-subclause (b) shall

(i)	accept the submission
or	
(ii)	negotiate

Subclause (3) then reverts to the provisions for ascertaining rates contained in 6th Edition Subclause (1) if agreement is not reached.

DC: Clause 52

Subclause (1) and (2)(a) and (b) follows the 7th.

Because there is no Bill of Quantities and associated rates, Clause 52(3) does not exist and a failure to agree through 52(d) leads to determination by the Employer's Representative of fair and reasonable valuation.

Highways: Clause 53 and 54

The provisions for obtaining quotations were intermixed with instructions and confirmation examined in Clause 53 above. If no agreement is reached then the provisions of Clause 54 "Valuation of Employer's Changes" apply. These broadly follow the ICE Contracts.

FIDIC: Clause 52

Subclause (1) follows the 5th but again, in Subclause (2) the Employer is to be consulted and varied rates agreed. Provision is made for the Engineer to use provisional rates or prices to value variations to maintain cashflow.

FIDIC includes a provision in Subclause (3) that 15% difference up or down in the Contract Price as a result of variations and changes in quantities can lead to an alteration in the Contractor's site an general overhead costs.

The daywork provisions are in Subclause (4) and are similar to Subclause (3) in the 5th.

New Engineering Contract: Clauses 61 to 64

The New Engineering Contract combines the procedures for dealing with Variations (now called Compensation Events) ordered by the Project Manager and matters perceived by the Contractor to be a variation. These are not separated to match the division in the ICE based contract.

There are four processes which are Notification, Quotation, Assessment and Implementation. These are examined below.

Although the Project Manager is also deemed to give early warning of an event it is only the Contractor who receives a sanction for not doing so as is particularised in "Notices and Claims Procedures" below.

Notification

Clause 61.1 envisages notification at the time of the Event which the Contractor undertakes and for which he is asked to provide a quotation.

Clause 61.2 is the basis for obtaining quotations for possible changes to the works which are not to be put into effect. It is obviously essential to be clear under which of these two clauses the compensation enquiry has been given.

In circumstances where the outcome is uncertain the Project Manager may provide criteria or assumption under Clause 61.6 which, if they are found to be incorrect, can lead to a reassessment.

Quotations

Clause 62.1 envisages that the Contractor may be required to submit multiple quotations for different means of dealing with a particular Compensation Event.

Clause 62.2 states that each quotation is to give the changes to the prices and particulars of any delay to the Completion Date, together with a revised programme. This is a laudable aim and envisages few changes to a scheme. Where the scale of the project is considerable and its uncertainty has lead to the choice of the Target Cost option it is unlikely that it will be manageable to provide a separate programme for each event. Many major projects are reverting to establishing the direct costs of individual events and then providing a revised programme to cater for Compensation Events over a period. It is only by these means that the Project Manager can avoid being overwhelmed by the flood of paper.

The timing of quotations is prescribed in Clause 62.3 as being three weeks from request by the Project Manager, who within two weeks will issue one of

- an instruction to submit a revised quotation;
- an acceptance of a quotation;
- a notification that a proposed instruction or a proposed changed decision will not be given; or
- a notification that he will be making his own assessment.

The Project Manager must give a reason for requiring a revised quotation (Clause 62.4) for which a further three weeks is permitted unless it is agreed that the time may be extended (Clause 62.5).

Assessment

Clause 63.1 provides that the changes to the prices are:

- the Actual Cost of the work already done;
- the forecast Actual Cost of the work not yet done; and
- the resulting Fee.

The possibility of reduction to the prices is within Clause 63.2.

The delay to the Completion Date is taken as being the date that the planned completion is later than that on the Accepted Programme. In practice this means the effect of inserting the duration of the event into the project management software running the Accepted Programme. The difference between this and the ICE based contracts is that extension and associated costs are linked and that there is no erosion of the Contractor's float.

The assessment is deemed to be based upon the Contractor having included for Contractor's risks that have a significant chance of occurring (Clause 63.5) such as weather delays occurring in the period that the varied work is being carried out.

The assessment is also based on the assumption that the Contractor acts promptly and competently to the Compensation Event. (This is obviously more important in the case of Project Manager assessment particularised below.)

The Project Manager assesses a Compensation Event (Clause 64.1) if:

- the Contractor has not submitted a required quotation and details of his assessment within the time allowed;
- the Project Manager decides that the Contractor has not assessed the compensation event correctly in a quotation and he does not instruct the Contractor to submit a revised quotation;
- when the Contractor submits quotations for a compensation event, he has not submitted a programme which this contract requires him to submit; or

- when the Contractor submits quotations for a Compensation Event the Project Manager has not accepted the Contractor's latest programme for one of the reasons stated in this contract.

The Project Manager uses his own programme to do this when:

- there is no Accepted Programme; or
- the Contractor has not submitted a revised programme for acceptance as required by this contract.

The Project Manager has a period of five weeks (being three weeks plus two weeks) starting from the need to produce the assessment.

Implementation

Implementation is not necessarily connected to instruction to carry out the Compensation Event. Implementation is notification that a quotation or Project Manager's assessment has been accepted and the total of the prices has been changed together with the planned Completion Date (Clause 65).

NOTICES AND CLAIMS PROCEDURES

5th Edition: Clause 52: Notice of Claims: 52 (4)

Subclause (4) is one of the most important in the Conditions of Contract, being the procedure to be following in conjunction with all claims under any clause of the Contract.

Subclause (4)(a): it is mandatory for the Contractor to give notice within 28 days of disagreement with any rates for variations formally notified to him. After this he will be time barred.

Subclause (4)(b): other than ordered variations covered in 4(a) the duty is to give notice "as soon as is reasonable after the happening of the events". The duty is laid on the Contractor to maintain adequate records to support any claim. This is irrespective of the Engineer requiring these under Subclause 4(c).

Subclause (4)(c): despite the fact that instructions to keep general or particular records, which could be agreed at the time, are without liability, many Engineers shy away from issuing such instructions. This is a pity as it introduces a further element of conflict; if the principle is admitted at a later date the records maintained by the Contractor may then be disputed. The Contractor is wise to submit his records on a regular basis with a covering letter so that, if totally ignored or perhaps even returned, they are introduced into the contractual record via the correspondence.

Subclause (4)(d): requires the provision of an interim account as soon as possible with "full and detailed particulars of the amount claimed...and the grounds upon which the claim is based." This is made mandatory by (4)(e) below. In practice this often means a full claim document. Without one the Engineer can claim that he does not have full and detailed particulars.

Subclause (4)(e): the importance of adhering to the procedure cannot be over-emphasised and the whole subclause is repeated below to highlight the severity of the provision.

> "If the Contractor fails to comply with the provisions of this clause in respect of any claim which he shall seek to make, then the Contractor shall be entitled to payment in respect thereof only to the extent that the Engineer has not been prevented from or substantially prejudiced by such failure in investigating the claim".

Subclause (4)(f) is the means by which the Engineer is obliged to make substantial interim payments on any part of the claim which has been substantiated to the satisfaction of the Engineer.

6th Edition: Clause 52
As 5th, but in Clause 52.4(b) there is now a longstop of 28 days to give notice.

7th Edition and DC: Clause 53
As 6th, but renumbered Clause 53.

Highways: Clause 55
The provisions are in Clause 55 'Procedure for Claims'. These broadly follow the ICE Conditions of Contract except with regard to timescales and the additional precondition of full time and cost predictions.

Clause 55.1 specifies 21 days for notice from commencement of the event or even earlier being when it ought first to have been discovered. This should be contrasted with the 28 days longstop after the event in the ICE Contracts.

Clause 55.3 "Substantiation of Claims" provides for an account with cost and contractual grounds to be provided. Long running events may have interim accounts and a final account 28 days after completion of the event.

FIDIC
The procedure is also in Clause 53 and is less prescriptive than the ICE Contracts. Subclause (1) states that notice shall be given 28 days after the beginning of the event giving rise to the claim.

Subclause (2) requires the Contractor to keep records and on receiving the notice in Subclause (1) requires the Engineer to inspect these and state what further records, if any, he requires. A more robust approach to establishing the facts relating to effects and quantum than in the other ICE based Contracts.

Subclause (3) requires an account within 28 days, interim accounts and a final account within 28 days of the end of the event.

Subclause (4) contains similar restrictions to the other ICE Contracts on failure to comply.

Subclause (5) regarding payment of claims again introduces consultation with the Employer before payment.

New Engineering Contract: Clauses 16, 61 and 63

The provisions are different in that there is the new concept of an early warning contained in Clause 16 which must be given by either the Project Manager or the Contractor as soon as they are aware of any matter which could increase the total of the prices.

A meeting must then be held at which all co-operate to either avoid or reduce the effect and decide who will take the actions.

The New Engineering Contract provisions follow the Notification, Quotation, Assessment and Implementation procedure described under "Valuation of Variation" with the following differences. Under Clause 61.3 the Contractor is permitted to notify a Compensation Event if he believes that the event is a Compensation Event, and

- it is less than two weeks since he became aware of the event; and
- the Project Manager has not notified the event to the Contractor.

Clause 61.4 gives the following powers to the Project Manager not to change the Completion Date or Prices if the Project Manager decides that an event notified by the Contractor:

- arises from a fault of the Contractor;
- has not happened and is not expected to happen;
- has no effect upon Actual Cost or Completion; or
- is not one of the compensation events stated in this Contract.

If the Project Manager decides otherwise, he instructs the Contractor to submit quotations for the event. Within either

- one week of the Contractor's notification; or
- a longer period to which the Contractor has agreed

the Project Manager notifies his decision to the Contractor or instructs him to submit quotations.

The Project Manager can also notify the Contractor if he decides that the Contractor did not give an early warning (Clause 61.5) and through 63.4 may assess the effect based on when the Contractor should have given such a warning.

The remainder of assessment and implementation is as particularised in "Valuation of Variations". In these cases a Project Manager's assessment is far more likely.

PLANT AND EQUIPMENT

5th Edition: Clause 53:

Plant Subclauses (1) to (5) deal with the vesting of plant owned by the Contractor in the Employer's name, the position with fixed plant and the requirement that hire agreements should provide for the Employer to hire the plant in the event of forfeiture.

Subclause (6) concerns the Contractor in the daily operation of the site as no plant may be removed without the Engineer's approval unless on outside hire.

Subclause (8) provides for the Engineer to dispose of plant or materials left on site. Subclauses (9) and (11) are disclaimers on liability for loss and injury and that vesting does not constitute approval.

Subclause (10) requires that the same conditions apply to Subcontractors.

6th Edition: Clause 53

Substantially rewritten, recognising that the Employer would only practically ever have a right to items owned by the Contractor as retention of title provisions frustrated the original provisions of the 5th.

7th Edition: Clause 54

Original Clauses 53 and 54 are combined into 54, presumably to return the numbering to the traditional sequence. The commensurate provisions are within Subclauses (1) to (3).

The specific difference is the omission of the caveat regarding removal of items which are "not immediately required". Hence, specialist plant intended to have two visits could theoretically not be removed at the end of the first.

DC: Clause 54

As 7th, but without the removal of the caveat described in the second paragraph.

Highways: Clause 53

Although differently structured the provisions are broadly as the 6th.

FIDIC: Clause 54
The provisions are in Clause 54. Subclause (1) introduces immovability of plant that corresponds with the 5th Edition. Subclauses (3) and (4) require the Employer to use his best endeavours to assist the Contractor in importing and re-exporting plant on completion. Subclause (5), combined with Subclause (7) in relation to subcontractors, requires that all hired plant brought on to site will include clauses entitling the Employer to hire it under the same terms and conditions to complete the works if the Contract is terminated.

New Engineering Contract: Clauses 70 and 72
Clause 70.2 recognises the reality of separate plant companies subcontracting and retention of title provisions by suppliers and states that "Whatever title" the Contractor has in Equipment, Plant and Materials passes to the Engineer when brought on to site and passes back to the Contractor when removed with the Project Manager's approval.

Clause 72 requires plant to be removed when it is no longer required unless the Project Manager instructs otherwise (to do so would involve payment). This is clearer than the 7th Edition of ICE.

VESTING OF OFF SITE ITEMS

5th Edition: Clause 54: Vesting of Goods and Material not on Site
Subclause (1) limits the application of this clause to materials which were listed in the Appendix to the Form of Tender. If they were not listed then there is no direct right to payment under the contract and it is to be hoped that the estimator has ensured that there are no Subcontractors whose quotes include off site payment without the items being in the Appendix. The further requirements are:

- that the materials are "substantially ready for incorporation into the works",
- and that they are the property of the Contractor by the means stated in Subclause (2).

Condition (a) is normally interpreted to include black steel, which can hardly be called substantially complete with the work of fabrication yet to do.

Subclause (2) sets out the procedure to be followed to obtain transfer of property and is one which the Contractor must follow and give instructions to his Subcontractors to follow. The Contractor must:

- provide documentary evidence that the goods are vested in the Contractor;
- mark goods with "Property of Employer" and with the name of the project or site;
- where relevant, such as in a stock steel store, set in a separate area;

- provide a schedule of the goods with value and an invitation to inspect.

Subclause (3) provides that, following approval by the Engineer, the goods become the property of the Employer and are not within the "ownership, control or disposition of the Contractor." Always provided that:

- the Employer pays the sums certified by the Engineer; and
- the Contractor remains responsible for all the loss, damage, handling and insurance.

Subclause (4) prevents any other person having a lien on the goods.

Subclause (5) covers the provisions in case of termination of the contract through Clause 63 and Subclause (6) requires that the Contractor ensures that all Subcontractors have the same vesting requirements incorporated into their subcontract. Hence the requirement on some contracts for proof of payment of the material supplier before vesting will be allowed. In Scotland vesting as described in the Contract does not apply, but the same effect is achieved through a Contract of Sale. When operating in Scotland the Contractor should discuss the format required with the Engineer at an early stage.

6th Edition: Clause 54
Structured as 5th but with the provision in Subclause (1) altered to permit the Engineer to require vesting of particular items.

DC: Clause 54
Provisions in Clause 54(4) onwards which match the 5th.

7th Edition: Clause 54
Provisions as 6th, but structured as DC.

Highways: Clause 26
Similar but differently structured provisions to those for the 6th are in Clause 26.

New Engineering Contract: Clauses 70 and 71
Clause 70.1 provides that title passes to the Engineer for Plant, Materials and Equipment which have been marked by the Supervisor for the contract and for which payment has been made. The entitlement to this cash flow enhancing procedure will be in the Works Information, otherwise there is no specific right to it.

Clause 71 makes the procedure subject to the Contractor demonstrating to the Project Manager that the goods are in accordance with the Works Information.

MEASUREMENT

5th, 6th, 7th Editions and DC Clause 55: Quantities, Correction of Errors

Provides that the Engineer may correct errors in the documents to be valued under Clause 52, but the Contractor cannot correct errors in his tender.

Highways: Clause 54

Clause 54 introduces the payment schedule linked to milestones of what is a lump sum contract.

FIDIC: Clause 55

Contains only Subclause 55(1) of the 5th and does not include provisions for correction of errors but this is not significant.

New Engineering Contract

Likely only to apply to Option B and not dealt with in core clauses.

MEASUREMENT AND VALUATION

5th Edition: Clause 56: Measurement – Valuation

Subclause (1) provides for the contract value to be ascertained and determined by admeasurement unless otherwise stated.

Subclause (2) provides for modifications to the rates where they have been rendered inapplicable by change in quantity. The Contractor should obviously be looking for such opportunities and they are discussed fully in Chapter 9.

It is generally accepted that the change in quantities must be of the order of 10% before this subclause applies, although in exceptional circumstances any change in quantities which leads to additional cost is covered. Such a rerate can be to recover spread monies and fixed overheads and in these cases decreases in quantities generally need to be greater than 10% before a rerate request is reasonable.

In some cases the nett rate is also no longer relevant and requires adjusting. One of the duties of the measurement team is to check all quantities to see if they qualify for rerating. Examples are given in the Claims Appendix. Increases of quantities must be viewed with an eye to increased resources and different methods required to complete the works in the allotted time on the programme. Care must be taken with re-rate applications for increased, as well as decreased, quantities on the same project. The Engineer may consider that the increased quantities lead to an over recovery of spread monies and reduced rate might be applied.

The general practice followed by most Contractors at tender stage is to try to spot under-measured items and increase their rates. If they are too ambitious this Clause can be used by the Engineer to claw back unreasonable gains.

Subclause (3) provides that, if the Contractor does not send a representative to measure the works when requested by the Engineer, the Engineer's measurement will be taken to be correct. Obviously it is most unwise for a Contractor not to send a representative.

6th Edition: Clause 56

As 5th except that simplified daywork procedures are incorporated at 56(4).

7th Edition: Clause 56

As 6th, with recognition in daywork procedures in 56(4) that FCEC is no more and schedule is issued by CECA.

FIDIC: Clause 56

Combines the 5th Edition provisions in Subclauses (1) and (3) in a single subclause and omits Subclause (2) regarding the ability to adjust rates that are not appropriate because of the change in quantity. This is a further risk that would suggest that contractors should keep as much temporary works costs out of rates as possible unless they consider an item under-measured. FIDIC Clause 52 does have the provision particularised earlier that a 15% change in total value leads to a review of prelims.

DC: Clause 56

Provisions are optional based on whether Bill of Quantities are included. They generally follow the 6th except that rerating through change of quantity is omitted.

Highways

As a Lump Sum Contract there are no provisions for remeasurement.

New Engineering Contract

Work will only be measured if Option B has been used and this is one of the lesser used options.

METHOD OF MEASUREMENT

5th, 6th and 7th Editions: Clause 57

Introduces into the documents the "Standard Method of Measurement of Civil Engineering Quantities". This clause will have had to be altered in the Contract documents for any other method to prevail.

In disputes arising from contradictions between the Standard Method and any other parts of the documents the Contractor should read carefully the differing legal opinions on which document takes precedence. It is impossible to generalise in this matter.

FIDIC

Presumption is that the Tender and Contract will contain a series of lump sums and that a breakdown of the sums will be submitted to enable progress payments to be made.

DC and HC

Payment is not linked in to measurement and, hence, there are no provisions.

New Engineering Contract

If Option B is used this will be specified in the Works Information.

CONTINGENCY SUMS

5th Edition: Clause 58: Provisional Sums and Prime Cost Items

Subclause (1) defines a Provisional Sum and the critical point is that it "may be used in whole or in part or not at all at the discretion of the Engineer." Hence there is no case for loss of overhead and profit recovery if it is not used. However, expenditure of a Provisional Sum must be considered against the approved Clause 14 programme.

Depending upon the itemisation of the Provisional Sum, if it affects the critical path then the Contractor would be entitled to consideration through Clause 13. Similarly if methods already approved are forced to change.

Subclause (2) defines the Prime Cost item which differs from the Provisional Sum in that it is an "item in the Contract" and will be used in whole or in part.

Subclause (3) states that design responsibility, if included, shall be passed to a Nominated Subcontractor. If the Subcontractor refuses to accept such responsibility then this would be valid grounds for objection under 59A(1).

Subclause (4) gives the Engineer the power to order the use of Prime Cost items either by nomination or for direct execution by the Contractor as a variation to be valued in accordance with Clause 52.

Subclause (5) ensures that anybody thrust upon the Contractor is to be deemed to be a Nominated Subcontractor.

Subclause (6) requires the Contractor to reveal all payments to a Nominated Subcontractor if requested to do so.

Subclause (7) provides that the Engineer may either order the Contractor to execute the works described in a Provisional Sum or to nominate a Subcontractor in a similar manner to his powers for the use of Prime Cost items.

6th and 7th Edition: Clause 58

The definitions of Provisional Sums, Prime Cost and Nominated Subcontractor in 58(1), (2) and (5) are deleted and particularised in Clause 1 definitions. Apart from reorganising the order the provisions remain the same.

DC: Clause 58

Simplified provisions which do not include Nominated Subcontractors. Design responsibility is not mentioned as it is inherent in this contract.

Highways

No provisions.

FIDIC: Clause 58

Simplified version of the 5th but does not include provisions for prime cost sums that will definitely be spent or protection against nomination of Subcontractors who will not accept the Contract design responsibilities.

New Engineering Contract

There are no specific provisions but an Employer who is uncertain of his final requirements will use an option that does not have a fixed cost.

NOMINATION

5th Edition: Clause 59A: Nominated Subcontractors

Subclause (1) allows the Contractor to object to any Nominated Subcontractor who broadly cannot or will not:

- meet the main contract specification or Contractor's Clause 14 programme;
- accept the full provisions of the main contract particularly with regard to damages;
- show proper and adequate insurances.
- accept the provisions of Clause 63 of the main contract.

Most cases of objection arise from the programme and the conditions of the main contract and the Contractor must ensure that the Subcontractor is totally back to back with him. This is done by using the FCEC Form of Subcontract. Refusal to accept this form of subcontract would normally be valid grounds for objection to the nomination. The Contractor could similarly not insist on anything more onerous.

In the broader wording of condition (a) in the document it is also possible to object to a Subcontractor whose finances are suspect. If the Contractor accepts the nomination without formal objection then, through Subclause (4), he is stuck with the Subcontractor as if he "had sublet the same in accordance with Clause 4". The only relief is in Subclause (6) where, for all Nominated Subcontractors, the Employer can only enforce damages or costs from an Engineer's or Arbitrator's award so far as the Contractor can get the monies from the Nominated Subcontractor.

This still leaves the Contractor with his own costs and that of other Subcontractors due to a breach of contract by a Nominated Subcontractor which is unsecured by the Employer.

Subclause (2) limits, and makes it mandatory for, the Engineer to do one of four things if the Contractor objects to a nomination:

1   nominate another Subcontractor who is subject to acceptance by the Contractor under subclause (1);
2   take the work out of the contract by variation for it to be performed by others;
3   subject to the Employer's consent (and the Contractor should ensure that he has a copy of this in writing) order the Contractor to enter into a subcontract on less onerous terms. Subclause (3) now applies and the **Contractor is protected for his own costs as well, if the Nominated Subcontractor defaults**;
4   arrange for the Contractor to execute the work.

On first sight it may seem obvious that the Contractor should always object to any nomination and decline to enter into a contract because of the greater protection and limited liability which come from Subclause (3) if the Engineer still instructs the Contractor to enter into such a subcontract. However, the Contractor must have valid grounds and cannot use his powers to decline to enter into a subcontract lightly, and through the provisions of 59A(2)(c) he can only do it for the reasons outlined in (a) to (d) of 59A(1).

Summary Advice for the 5th Edition
(a) The Contractor must scrutinise the quotation of the Nominated Subcontractor to ensure that it conforms to the documents (particularly with regard to services to be provided by the Main Contractor), his Clause 14 programme and approved method statements.

(b) He must also satisfy himself that the Subcontractor is financially sound before authorising the placing of a subcontract.

5th Edition: Clause 59B: Forfeiture or Termination of the Nominated Subcontractor

The Contractor would be most unwise to terminate a Nominated Subcontractor without either the Employer's consent or the Engineer's direction. In practice this must mean that as soon as a notice is given under Clause 17 of the subcontract for the Subcontractor to do something in a given time or face specified consequences, the Contractor should send a copy to the Engineer and discuss the matter with him.

If termination is eventually made with the consent of the Engineer, then through Subclause (4)(b) the Contractor is entitled to a paid extension of time together with his costs. Subclause (4)(c) then allows the Employer to claim his additional costs from the Contractor but is limited through Subclause (6) to be entitled to that which the Contractor can get after due "proceedings" from the consent of the Employer or the Engineer's direction.

6th and 7th Editions: Clause 59

Clauses A and B are now combined. The omission in the opening paragraph of 59(1) of a link to 2(c) reflects the deletion of this provision below. In Subclause 1(d) the Contractor can now ask for security, normally a performance bond, as a further precondition.

Subclause (2): This Clause now not only governs the action of the Engineer in the event that the Contractor has a valid objection to a subcontractor nominated by the Engineer but also covers the situation where a re-nomination is necessary following a valid termination of a Nominated Subcontract.

The Clause is substantially as previous Clause 59A(2) but omitting subclause (2)(c) so that there is no provision for the employment of a Nominated Subcontractor on terms that are in conflict with the provisions of Clause 59(1). Subclause 59(2)(d) has been added to provide specifically an option for the Contractor himself to obtain a quotation from a direct Subcontractor for carrying out the work.

The provisions originally in Subclause 59A(3) linked to reduced liability are also deleted.

FIDIC: Clause 59

The Contractor is afforded similar rights and recognised bases to object to Nominated Suppliers or Subcontractors to those within the 5th. The Employer does not have the right to overrule the Contractor and force him to use the Subcontractor. UK case law requires, in this event, that there be re-nomination.

The Employer is permitted to pay the Nominated Subcontractor directly and offset where the Contractor does not pay in accordance with a certificate without due course.

DC and Highways
No provision for nomination.

New Engineering Contract
There are no provisions for Nominated Subcontractors. If there are Named Subcontractors in the Works Information they are to be treated as Domestic Subcontractors. If they cannot be brought into line then the Contractor should approach the Project Manager for direction. (See Subcontracting earlier and the need for compatible conditions).

Back to back contractual liability with a named Subcontractor should be achieved through the provisions of Clause 26(2) and in particular Subclause (3) which requires compatibility in obligations. The Project Manager will have no sympathy with more onerous conditions than the NEC Subcontract (see Subcontract Appendix).

PAYMENT

5th Edition: Clause 60: Payments
Subclause (1) gives the format for the Contractor's monthly statements namely:

- value of Permanent Works executed;
- materials on site and their value;
- list of vested material in accordance with Clause 54;
- value against any separate temporary works items and any claim or additional payment requested.

Provision is also made for Nominated Subcontractors to be listed separately.

Subclause (2) contains the important provisions that "within 28 days of delivery to the Engineer or Engineer's Representative the Engineer should certify and the Employer shall pay."

If the Engineer does not request a valuation submission timetable the Contractor should arrange one. It will be the list of "specified dates" which are to be included in any subcontract let under the CECA Blue Form of Subcontract. The dates themselves will have been arranged so that the Contractor's cost period end is just more than 28 days from the specified date. The money will then be in the bank at the end of the period when most banks charge or pay interest, and then Subcontractors are paid seven days later in the next month.

The remainder of the subclause deals with the Engineer's power to amend the submission before certification.

Subclause (3) requires the Contractor to submit a final account with supporting documentation within three months of issue of the Maintenance Certificate. Within a further three months the Engineer shall issue a Final Certificate. This latter period is also subject to the Engineer receiving "all information reasonably required for its verification", so it behoves the Contractor to be complete and prompt in reply to any queries.

Subclauses (4) and (5) deal with the payment of retention. The need for prompt application for Clause 48 Certificates and the release of retention has been discussed under Clause 48.

Subclause (6) introduces the Contractor's right to interest on overdue payments.

In instances where modifications are made to exclude this clause it still does not remove the entitlement to finance costs providing the Contractor is actually borrowing.

Subclause (7) allows for the Engineer to modify all interim amounts certified other than those to Nominated Subcontractors.

6th Edition and DC: Clause 60

Subclause (1) modified to recognise monthly intervals rather than the end of the month for interim payments.

New Subclause (3) has split off from Subclause (2) and makes it clear that minimum certificate provisions only apply up to the date of substantial completion.

Subclause (5), previously (4), does not now provide for specified retention rates and these are now specifically to be stated in the Appendix to the Form of Tender.

Subclause (6): Substantially as previous Clause 60(5). Reference is now made to half of the relevant proportion of the retention money deducted instead of 1.5% of the amount due to the Contractor. Retentions are released when Sections or parts of the Works are completed to be in proportion to value and to be included in certificate next issued.

Subclause (7): the previous Clause 60(6) has been expanded to make it clear that interest is payable for each day a payment is overdue and that the interest is to be compounded monthly so that interest is added each month to the interest already due to be paid on an overdue payment.

The wording of the Clause has also been revised to refer to the bank specified in the Appendix to the Form of Tender. An addition has been made so that the English and Scottish procedures for the award of interest on Certificates corrected by an Arbitrator are the same.

7th Edition: Clause 60

As 6th, except modifications at 60(2) to 25 days, 60(6)(c) to 10 days, insertion in 60(7)(b) of reference to Adjudicator's findings and insertion in 60(1) of requirements on the Employer to provide reasons for withholding certified monies; all to be compatible with the Housing Grants Act.

Highways: Clauses 58, 59 and 63

The equivalent provisions are within three clauses, being 58 Milestone Achievement; 59 Amendments to Milestone Achievements; and 63 Withholding from Certificates.

Clause 58.1 provides for payment of milestones achieved early in the appropriate monthly account. The postponement of payment until milestones are achieved is within 58.2.

In the latter case, the Employer's Agent may issue a new payment schedule. However, the Contractor will still get paid when he achieves milestones, even if they are ahead of the schedule.

Clause 60.1 provides for monthly payments of milestones achieved, claims, residual sums from the previous payment and sums due from vesting of plant and materials not on site.

The provisions for withholding sums otherwise due above against improper work are within Clause 63 and mirror ICE Condition 60(8). These will eventually result in "residual payments" when corrected.

Clause 60.2.1 requires the Employer's Agent to certify within five days of receipt of monthly details and 60.2.2 within 14 days of payment being due for residual payments.

Retentions are set at 5 and 2.5%.

The final account is termed a Reconciliation Account. The timetable is:

- submission six months after Taking Over Certificate;
- Employer's Agent period of review maximum three months to issue of Reconciliation Certificate;
- final payment within 28 days.

No specific period for acceptance by the Contractor is stated so this must revert back to the procedure for claims and the 21 days in Clause 55.1. In these circumstances it is unlikely that the caveats in 55.4 would apply and the 21 days would probably hold.

Separate payment provisions are established in 60.5 for work within the maintenance period so that this does not prevent completion of the Reconciliation Account.

Provisions are included for retentions to be released with a Retention Bond.

Interest provisions are as ICE except that interest provisions are to be specified in the Contract documents rather than the Conditions of Contract.

FIDIC: Clause 60

After basically following the 5th for Subclauses (1) and (2), FIDIC now follows a different direction. Subclause (3) details the release of half of the retention after the Taking Over Certificate and the other half after the Defects Liability period. Subclause (4) gives the Engineer the right to correct interim certificates.

There is then a two-stage procedure: it commences with Subclause (5) and a "Statement at Completion" 84 days after the Taking Over Certificate. This is then followed in Subclause (6) with a "Final Statement" which is to be presented 56 days after the Defects Correction Certificate and is to be agreed with the Engineer.

With the Final Statement the Contractor is to provide a written discharge as described in Subclause (7). Twenty eight days after receipt of both of these the Engineer is to certify as particularised in Subclause (8).

Subclause (10) provides for payment of Interim Certificates 28 days after delivery to the Employer and 56 days for the Final Payment. Thus, under FIDIC the interim payment period is doubled from standard ICE provisions.

New Engineering Contract: Clauses 50 and 51

In all the other contracts the payment of interim sums is initiated by the Contractor making an application and it being assessed on behalf of the Employer. The New Engineering Contract is different in that Clause 50.1 places the responsibility on the Project Manager to make the assessment no later than the assessment intervals leading off from the starting date. Payments are also to be made at completion of the Whole of the Works, when the Supervisor issues the Defects Certificate. He is to consider any submission made by the Contractor and is to inform him how the sum certified is built up.

Clause 50.5 permits corrections of wrongly assessed amounts in subsequent payments.

The Project Manager is to certify within one week of the assessment date (Clause 51.1) and payment is to be made within four weeks of the assessment date. Compound interest is to be paid in:

- late payments from the date due (Clause 51.2);
- late certificates (Clause 51.4); and
- corrected values of Compensation Event (Clause 51.3).

MAINTENANCE CERTIFICATE

5th Edition: Clause 61

Upon expiry of the maintenance period referred to in Clause 48 and when all defects have been made good, this is issued. Stated not to relieve either party of their obligations.

6th and 7th Editions: Clause 61

As 5th but now called Defects Correction Period.

DC: Clause 61

As 6th and 7th but Clause 61(3) also requires draft Maintenance Manuals and as constructed drawings to have been submitted and approved by the Employer and three copies issued to him. In the UK the CDM Regulations require the Health and Safety File which has all this detail in it to be available far earlier.

Highways: Clauses 51 and 64

The provisions are split between Clause 51(8) which provides that the Designer (employed by the Contractor) certify in a prescribed form that the Contractor has complied with all the obligations under the Contract before the Employer's Agent issues the certificate, and Clause 64 which makes the certificate without prejudice to continuing warranties.

FIDIC: Clauses 61 and 62

Clause 61.1: only the Defects Liability Certificate is deemed to constitute approval of the works. Clause 62 follows 6th and 7th.

New Engineering Contract

The requirements are at Clause 11 and Subclause 15:

> (15) The Defects Certificate is either a list of Defects notified before the defects date which the Contractor has not corrected or, if there are no such Defects, a statement that there are none.

DISPUTE RESOLUTION

5th, 6th, 7th Editions and DC: Clause 66

Clause 66 was amended in March 1998 (where the Contracts or their equivalents are used in the UK) in the 5th, 6th and DC to cover the introduction of the Housing Grants, Construction and Regeneration Act 1996 (HGCRA). The provisions of the Clause now comply with that Act and this is as included in the 7th; making all four compatible.

The Engineer's decision has been retained as a condition precedent to all dispute resolution procedures under the ICE Conditions. As HGCRA provides that adjudication can take place "at any time", a dispute under Clause 66 can now only come into existence after the Engineer has given his decision (or the time allowed for that purpose has expired). This should ensure that, once a dispute has been notified, the issues will already have been properly defined before any reference to adjudication can take place.

Conciliation has been retained as an option open to the parties and effectively is introduced into the 5th.

Finally, in the event of a reference to arbitration, the parties have a choice between the ICE Arbitration Procedure (1997) and the new Construction Industry Model Arbitration Rules. That choice is made in the Appendix to the Form of Tender before tenders are invited.

Thus, within the UK there is procedure introduced in 66(2) which requires a notice of dissatisfaction and referral to the Engineer with a timescale of one month. On expiry, or an adverse decision, a Notice of Dispute can be served by the Contractor or Employer on the Engineer (which must fully particularise the dispute).

There is a significant reduction in the status of the Engineer's Decision in the UK. Previously, it took three months and was final after a further three months if the disputing party did not request Arbitration which still applies when the Contract is used outside the UK.

The new Notice of Dispute is necessary before embarking on either Conciliation, Adjudication or Arbitration. It is important to realise that the Notice of Dispute in itself does not set any clock ticking with regard to timescales. Each of the subsequent options has its own notice and a matter can be "parked" indefinitely before the final account.

The disputing party (hereinafter assumed to be the Contractor) has a choice of routes. Assuming no settlement at intermediate stages he may go via Conciliation to Adjudication to Arbitration or directly to Adjudication and then to Arbitration. Although the defendant is unlikely to agree, being in possession of a legally binding award, it is also possible to go from Adjudication to Conciliation to Arbitration. All of these require a notice and then the timescales start.

Conciliation under Subclause (5)(a) is by agreement. The procedure is relatively cheap and permits the parties to retain control of the dispute and the settlement.

There may be an advantage on the bigger delay and disruption claims to go this way first because the prescriptive times in the Adjudication process could lead to a lottery.

There is now no time limit on the process because it is not being fitted in after Arbitration has been notified. It is a separate standard, the only time proviso is appointment within 14 days.

The provision in Subclause (5)(b) reflects the ICE Conciliation Procedure where, 28 days after the Conciliator's written recommendation it becomes a debt due if not challenged.

This new procedure within Clause 66(5)(b) requires the challenge to be either a Notice of Adjudication or a Notice to Refer to Arbitration.

Subclause (6) introduces Adjudication which can occur at any time after a Notice of Dispute by the service of a "Notice of Adjudication".

The timetable is then fast, being seven days for appointment and 28 days extendable by 14 day on the applicant's consent before the award.

Subclause (9)(b) provides that an Adjudicator's decision is fully final after three months, unless a Notice to Refer is served, and the right to Arbitration disappears.

Where these Contracts or their derivatives are used outside the UK the Adjudication provisions are not included and Arbitration must be notified 3 months after the Engineer's decision.

Highways: Clause 68: Disputes Resolution

Dispute is defined in Clause 1.1 as any difference between the Employer and the Contractor. There is no requirement for the Employer's Agent to rule upon it.

There is no provision for Conciliation.

All disputes must be referred firstly to Adjudication by the service of a "Notice". Timescales for Adjudication process are then as the ICE, being set by the provisions of the Housing Grants Act.

Interestingly, there is no time limit on when either party can issue a "Notice to Refer" to Arbitration, which is contained within the ICE Conditions of Contract.

FIDIC: Clause 67

Subclause (1) provides that disputes are to be referred to the Engineer in writing with a copy to the other party. The Engineer is to rule within 84 days. If either party is dissatisfied or the Engineer fails to give a decision in 84 days then within 70 days of the Decision, or 70 days from when it ought to have been given, either party may give notice of intention to commence arbitration proceedings. If no such application is made the Decision is binding.

Subclause (3) introduces "Rules of Conciliation and Arbitration of the International Chamber of Commence" which also governs appointments.

Arbitration can be commenced before the completion of the works and neither side need stick to the evidence and arguments put to the Engineer.

New Engineering Contract: Clause 90

The New Engineering Contract brings non-UK users the benefits of Adjudication which is currently available in all contracts within the UK through the Housing Grants Act, although the 28-day timescale need not apply. The principles of the Dispute Procedure are simple with two stages, being Adjudication and Review by a Tribunal. The Employer is given added powers in that he may terminate the Contract if the Contractor does not perform.

Commencing with Adjudication, Clause 90.1 provides a table which indicates that only the Contractor can challenge actions or lack of action by the Project Manager or Supervisor. The timescales are defined as being four weeks from the event to notification and the submission to the Adjudicator to be between two and four weeks thereafter. Any other matter may be referred to the Adjudicator by either party and this must be between two and four weeks from the notification.

Clause 90.2 requires that the Adjudicator give his decision to the parties and the Project Manager with reasons within the time provided by the Contract. Until this time, or there is settlement, matters proceed as if there was no dispute. The decision is final unless taken to Tribunal.

The period for the Adjudication can be eight weeks because Clause 91.1 provides for a four week period to provide further information and four weeks maximum for the decision. Any communications between a party and the

Adjudicator are also copied to the other party and the decision is operated in the same way as a Compensation Event (Clause 92.1).

The provisions for referral to a Tribunal are simple and cannot be sensibly paraphrased, and are therefore inserted below:

93.1 If after the Adjudicator

- notifies his decision or
- fails to do so

within the time provided by this contract a Party is dissatisfied, that Party notifies the other Party of his intention to refer the matter which he disputes to the Tribunal. It is not referable to the Tribunal unless the dissatisfied Party notifies his intention within four weeks of

- notification of the Adjudicator's decision or
- the time provided by this contract for this notification if the Adjudicator fails to notify his decision within that time

whichever is the earlier. The tribunal proceedings are not started before Completion of the whole of the works or earlier termination.

93.2 The Tribunal settles the dispute referred to it. Its powers include the power to review and revise any decision of the Adjudicator and any action or inaction of the Project Manager or the Supervisor related to the dispute. A Party is not limited in the tribunal proceedings to the information, evidence or arguments put to the Adjudicator.

NOTICES AND COMMUNICATIONS

5th Edition: Clause 68: Notices
Notices to Employer's and Contractor's registered office or main place of business.

6th Edition and DC: Clause 68
Additional Clause 1(6) defines communications by all means, arguably not e-mail.

7th Edition: Clause 68
As 6th, but adds provision for Engineer to specify where notices in writing to him are to be sent and clause 1 (6) now expanded to definitely include email.

FIDIC: Clause 68
Anticipates that address for notices will be in Part II but means of communication does not include email.

Highways: Clause 69
Provides greater specific detail including when notices are deemed to have been received.

New Engineering Contract: Clause 13
Only the NEC is specific about the timing of communications, Clause 13 and the period in which the Project Manager has to reply (which will be in the Contract Date part of the Works Information). This is a very useful innovation which provides a certainty as to when notices will be replied to, which is not in any other form of contract where a reasonable but disputable time will be deemed to apply.

Clause 13.2: provides that notice takes effect when received at last notified address of the recipient or if none is notified that stated in the Contract Data.

Clause 13.1 permits notice to be in a form that can be read, copied and recorded which includes email.

TAX MATTERS

5th, 6th, 7th Editions and DC: Clauses 69 and 70: Tax Matters
These clauses protect the Contractor in the event of tax fluctuations or changes in the course of the contract. Currency and exchange rates not included.

FIDIC: Clause 70
Provisions are confined to Clause 70. Clause 70.1 provides for fluctuations from stated labour and material rates anticipated to be in Part II. Clause 70.2 permits cost changes from legislation enacted 28 days prior to the Tender to be included.

Currency restrictions imposed 28 days prior to the latest date for submission are reimbursable. Where payment is in a particular currency or mix of currencies the payments in any particular currency are not adjusted to reflect rate variations.

Highways
No protection from tax fluctuations. Only recovery will be if the tax would not otherwise have been incurred but for an Employer's change.

New Engineering Contract
No specific inclusion in the Conditions.

Index